WORKBOOK

D0005042

COREQUISITE SUPPORT MODULES FOR QUANTITATIVE REASONING OR LIBERAL ARTS MATH

Corequisite Support Faculty Team

Pearson

23 2023

ISBN-13: 978-0-13-575396-5
ISBN-10: 0-13-575396-1

Contents

Copyright © 2020 Pearson Education, Inc.

1.5 Proportions
1. Solve proportions.
2. Solve applications involving proportions.

1.6 Measurement
1. Identify U.S. units of length, weight, and capacity.
2. Perform unit conversions among U.S. units (including mixed units).
3. Identify metric units of length, mass, and capacity.
4. Perform unit conversions among metric units.
5. Convert between U.S. and metric units.
6. Solve applications involving units of measurement.

1.7 Real Numbers
1. Classify sets of numbers.
2. Find square roots.
3. Approximate square roots.
4. Use the properties of real numbers.
5. Use the order of operations with real numbers (including grouping symbols).
6. Solve applications involving real numbers.

Module 2: Linear Equations and Inequalities; Formulas CS-99
2.1 Algebraic Expressions
1. Evaluate algebraic expressions.
2. Combine like terms.
3. Simplify algebraic expressions.
4. Translate English phrases into algebraic expressions.

2.2 Linear Equations in One Variable
1. Distinguish between expressions and equations.
2. Solve linear equations in one variable using the addition property of equality.
3. Solve linear equations in one variable using the multiplication property of equality.
4. Solve linear equations in one variable using both properties of equality.
5. Translate sentences into equations.
6. Solve applications involving linear equations in one variable.

2.3 Linear Inequalities in One Variable
1. Write inequality statements using real numbers and inequality symbols.
2. Graph linear inequalities in one variable on a number line.
3. Write solutions to inequalities in set-builder notation.
4. Write solutions to inequalities in interval notation.
5. Solve linear inequalities in one variable.
6. Translate sentences into linear inequalities in one variable.
7. Solve applications involving linear inequalities in one variable.

2.4 Formulas
1. Solve a formula for a specific variable.
2. Find the perimeter of a figure.
3. Find the circumference of a circle.
4. Find the area of a figure.
5. Find the volume of a figure.
6. Solve applications involving distance, rate, and time.

Module 3: Graphing Linear Equations in Two Variables CS-145
3.1 The Rectangular Coordinate System
1. Write ordered pairs.
2. Plot points in the rectangular coordinate system.

3. Complete a table of values of ordered pair solutions for a linear equation in two variables.
4. Graph linear equations in two variables using a table of values.

3.2 Intercepts
1. Find the intercepts of a line.
2. Graph a linear equation in two variables given its intercepts.

3.3 Slope
1. Find the slope of a line using the slope formula.
2. Find the slope of a line given its graph.
3. Graph a line given its equation in slope-intercept form.
4. Graph a line given one point on the line and the slope.
5. Graph vertical lines.
6. Graph horizontal lines.
7. Use slope with parallel and perpendicular lines.
8. Interpret slope as a rate of change.

3.4 Equations of Lines
1. Find the slope of a line given its equation.
2. Write the slope-intercept form of a line.
3. Write the equation of a line given the slope and a point on the line.
4. Write the equation of a line through two given points.

Module 4: Exponents, Polynomials, and Quadratic Models CS-181
4.1 Exponential Expressions and Rules for Exponents
1. Evaluate exponential expressions with positive exponents.
2. Use the product rule for exponents.
3. Use the power rules for exponents.
4. Use the quotient rule for exponents.
5. Evaluate exponential expressions with integer exponents.
6. Simplify exponential expressions using the rules for exponents.

4.2 Scientific Notation
1. Convert between scientific and standard notation.
2. Perform calculations involving scientific notation.
3. Solve applications involving scientific notation.

4.3 Polynomial Expressions
1. Identify parts of a polynomial (coefficient, term, degree, factor, constant).
2. Classify polynomials.
3. Evaluate polynomial expressions.
4. Add polynomials.
5. Subtract polynomials.
6. Multiply monomials.
7. Multiply a monomial and a polynomial.
8. Multiply polynomials.
9. Multiply the sum and difference of two terms.
10. Square binomials.

4.4 Factoring
1. Factor out the GCF of a polynomial.
2. Factor trinomials with a leading coefficient of 1.
3. Factor trinomials with a leading coefficient other than 1.
4. Factor polynomials by grouping.
5. Factor a difference of squares.
6. Factor the sum or difference of two cubes.
7. Factor polynomials completely.

4.5 Quadratic Equations and Models
1. Solve quadratic equations by factoring.
2. Solve quadratic equations by the square root property.
3. Solve quadratic equations by completing the square.
4. Solve quadratic equations by using the quadratic formula.
5. Distinguish between linear and quadratic models in real world situations.
6. Solve applications involving quadratic equations and models.

Module 5: Financial Math   CS-245
5.1 Percent Applications: Sales Tax, Commission, Discount
1. Calculate sales tax, total price, and sale price.
2. Calculate commission.
3. Calculate tips.
4. Calculate original price, discount, total cost, tax, and markup.
5.2 Simple Interest
1. Compute simple interest.
2. Find principal, interest rate, or time in the simple interest formula.
3. Solve applications involving simple interest.
5.3 Compound Interest
1. Compute compound interest.
2. Find principal, interest rate, or time in the compound interest formula.
3. Solve applications involving compound interest.

Module 6: Introduction to Functions   CS-265
6.1 Relations and Functions
1. Identify relations and functions.
2. Evaluate functions using function notation.
3. Find the domain and range of a function.
4. Use the vertical line test to determine if a graph is a function.
6.2 Linear Functions
1. Identify linear functions.
2. Evaluate linear functions.
3. Interpret the graph of a linear function (domain, range, slope, intercepts).
4. Solve applications that involve linear functions as models.

Module 7: Introduction to Statistics   CS-281
7.1 Data Displays
1. Interpret and draw line graphs.
2. Interpret and draw bar graphs.
3. Construct frequency distributions and relative frequency distributions for a data set.
4. Interpret and draw histograms.
5. Interpret and draw circle graphs.
6. Interpret and build scatterplots.
7.2 Measures of Center
1. Find the mean of a data set.
2. Find the weighted mean.
3. Find the median of a data set.
4. Find the mode of a data set.
5. Find the range and midrange of a data set.

Module 8: Introduction to Probability   CS-303
8.1 Counting Techniques
1. Evaluate factorial expressions.
2. Use counting techniques.
8.2 Introduction to Probability
1. Identify sample spaces, outcomes, and events.
2. Find the probability of an event.
3. Use tree diagrams to find sample spaces and compute probabilities.

Module 9: Sets and Logic   CS-313
9.1 Sets
1. Define terminology associated with sets.
2. Describe the members of a set using various notations.
3. Find subsets of a set.
4. Find equivalent sets.
5. Find the union of two sets.
6. Find the intersection of two sets.
7. Find the complement of a set.
9.2 Reasoning, Arguments, and Statements
1. Distinguish between inductive and deductive reasoning.
2. Distinguish between valid and invalid arguments.
3. Identify types of statements.

CRITICAL THINKING
The Critical Thinking worksheets provide students with extended exercises or mini activities to help them make conceptual connections. There are word problems to help with math-reading connections. There is at least one critical-thinking exercise per topic, and some that help students make connections between topics.

ACTIVITIES

This section contains one extended activity per module. These activities could be used for group-work and are designed to help students connect the module review content with what they are learning in their gateway course.

Welcome to the Workbook for the Corequisite Support Modules for Quantitative Reasoning or Liberal Arts Math

What are the Corequisite Support Modules?

Corequisite Support Modules for Quantitative Reasoning or Liberal Arts Math provide targeted developmental review to support the concepts in a Quantitative Reasoning or Liberal Arts Math corequisite course. Topics were selected based on syllabi collected from corequisite courses around the country.

Module instruction and practice is available within this workbook and MyLab Math. Content is flexible, allowing instructors to pick and choose the corequisite review objectives they need, when they need it. Topics are not heavily styled in approach so that they work well in conjunction with *any* Quantitative Reasoning or Liberal Arts Math credit-level materials.

The material was assembled by the **Corequisite Support Faculty Team** - instructors who were tasked with implementing coreqs at their own schools. They partnered with Pearson to compile the MyLab Math course and this workbook.

This Corequisite Workbook supports different class formats

Often with corequisite implementation, developmental-level students enrolling in a corequisite course benefit from hands-on practice or active learning in the classroom when it comes to reviewing developmental topics. The Workbook is ideal for instructors looking for resources to use in the classroom to get students putting pencil to paper.

The Workbook is comprised of three key parts: 1) Core Skills, 2) Critical Thinking, and 3) Activities. Instructors can opt to use each type of worksheet as appropriate for their corequisite support classroom—those who want an emphasis on skills practice for particular topics may focus on Core Skills, while those who want to incorporate conceptual understanding or active learning may also pull from Critical Thinking and Activities. The Critical Thinking section is designed to help solidify the concepts, and often demonstrate how they connect to one another. The Activities are designed to help students connect the developmental material to the quantitative reasoning or liberal arts math concepts.

Use it alone, or with a corresponding MyLab course

This Workbook can be adopted by itself to provide students with the foundational skills needed for success in a corequisite Quantitative Reasoning or Liberal Arts Math support course, or it can be paired with the MyLab course for additional practice and resources. The MyLab course offers premade homework assignments, interactive exercises, short instructional videos for objectives, and digital material to support a growth mindset.

The workbook and the MyLab Modules cover the same nine Modules' worth of developmental math content—content that has been developed and curated *specifically* to support Quantitative Reasoning or Liberal Arts Math.

1.1.1　Simplify fractions.

The number $\dfrac{3}{7}$ is a fraction, used to represent 3 parts of a whole that has been divided into 7 equal parts.

The number on the top of the fraction is called the **numerator**, and the number on the bottom of the fraction is called the **denominator**.

$$\frac{\text{numerator}}{\text{denominator}}$$

If the numerator and denominator of a fraction do not have any common factors other than 1, the fraction is said to be in **lowest terms**.

To simplify a fraction to lowest terms, begin by finding the prime factorization of both the numerator and denominator. Then divide the numerator and the denominator by their common factors.

$$\frac{105}{350}=\frac{3\cdot5\cdot7}{2\cdot5\cdot5\cdot7}=\frac{3\cdot\cancel{5}\cdot\cancel{7}}{2\cdot\cancel{5}\cdot5\cdot\cancel{7}}=\frac{3}{10}$$

Example	**Exercise 1**
Simplify $\dfrac{10}{18}$ to lowest terms.	Simplify $\dfrac{42}{49}$ to lowest terms.
SOLUTION: Begin by factoring the numerator and denominator. The numerator and denominator have a common factor of 2 that can be divided out of each. $$\frac{10}{18}=\frac{2\cdot5}{2\cdot3\cdot3}$$ $$=\frac{\cancel{2}\cdot5}{\cancel{2}\cdot3\cdot3}$$ $$=\frac{5}{9}$$	

Exercise 2

Simplify $\dfrac{8}{40}$ to lowest terms.

Exercise 3

Simplify $\dfrac{168}{378}$ to lowest terms.

Exercise 4

Simplify $\dfrac{45}{77}$ to lowest terms.

Exercise 5

Simplify $\dfrac{91}{119}$ to lowest terms.

1.1.2   Plot fractions on a number line.

To plot a fraction on a number line, first determine which two consecutive integers the fraction is between.

Divide that interval into equally spaced sections as determined by the denominator of the fraction. For example, if the denominator is 6 the interval can be divided into sixths by placing 5 equally spaced marks between the two integers.

Use the numerator to determine which of those marks will be used to plot the fraction.

For example, suppose you wanted to plot the fraction $\frac{2}{5}$.

The fraction is between 0 and 1.

To divide that interval into fifths, place 4 equally spaced marks between 0 and 1.

Those marks represent $\frac{1}{5}, \frac{2}{5}, \frac{3}{5}$, and $\frac{4}{5}$.

Plot the fraction at $\frac{2}{5}$ on the number line.

Example	Exercise 1
Plot $\frac{3}{4}$ on a number line.	Plot $\frac{2}{7}$ on a number line.
SOLUTION:	
The fraction $\frac{3}{4}$ is between 0 and 1.	
Divide that interval into fourths on the number line, and plot the fraction.	

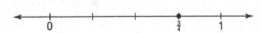

Exercise 2

Plot $-\dfrac{5}{8}$ on a number line.

Exercise 3

Plot $\dfrac{8}{5}$ on a number line.

Exercise 4

Plot $-\dfrac{22}{9}$ on a number line.

Exercise 5

Plot $8\dfrac{2}{3}$ on a number line.

1.1.3 Add or subtract like fractions.

To add two fractions that have the same denominator, add the numerators and place the sum over the common denominator. Simplify the resulting fraction to lowest terms.

$$\frac{5}{12} + \frac{11}{12} = \frac{16}{12}$$

$$= \frac{\overset{4}{\cancel{16}}}{\underset{3}{\cancel{12}}}$$

$$= \frac{4}{3}$$

To subtract a fraction from another fraction that has the same denominator, subtract the numerators and place the difference over the common denominator. Simplify the resulting fraction to lowest terms.

$$\frac{7}{15} - \frac{4}{15} = \frac{3}{15}$$

$$= \frac{\overset{1}{\cancel{3}}}{\underset{5}{\cancel{15}}}$$

$$= \frac{1}{5}$$

Example	**Exercise 1**
Simplify $\dfrac{5}{16} + \dfrac{7}{16}$.	Simplify $\dfrac{3}{10} + \dfrac{1}{10}$.
SOLUTION: Add the numerators and place the sum over the common denominator. Simplify the resulting fraction to lowest terms. $$\frac{5}{16} + \frac{7}{16} = \frac{12}{16}$$ $$\frac{3}{4}$$	

Exercise 2

Simplify $\dfrac{19}{24} - \dfrac{11}{24}$.

Exercise 3

Simplify $\dfrac{9}{25} - \dfrac{21}{25}$.

Exercise 4

Simplify $\dfrac{17}{100} + \dfrac{31}{100}$.

Exercise 5

Simplify $\dfrac{13}{56} - \dfrac{41}{56}$.

1.1.4   Find the least common denominator of a list of fractions.

The **least common denominator** (**LCD**) of a list of fractions is the smallest whole number that each denominator can divide into evenly.
In other words, this is the smallest whole number that is a multiple of each denominator.
For example, if two denominators are 4 and 6, their least common denominator would be 12 because 12 is the smallest multiple of both 4 and 6.

Finding the Least Common Denominator
- Find the prime factorization of each denominator.
- Find the common factors of the denominators.
- Multiply the common factors by the remaining factors of the denominators.

Suppose you needed to find the least common denominator if $\dfrac{3}{8}$ and $\dfrac{7}{10}$.

Find the prime factorization of each denominator: $8 = 2 \cdot 2 \cdot 2$ and $10 = 2 \cdot 5$.
The common factor is 2. Multiply the common factor by the remaining factors (2, 2, and 5).
Least Common Denominator: $2 \cdot 2 \cdot 2 \cdot 5 = 40$

Another approach that can be used is to start listing multiples of each denominator until you find one that is a common multiple of each denominator.

Suppose you needed to find the least common denominator if $\dfrac{3}{8}$ and $\dfrac{7}{10}$.

Multiples of 8: 8, 16, 24, 32, 40, …   Multiples of 10: 10, 20, 30, 40, …
The least common denominator would be the first number to appear in both lists, which is 40.

Example	Exercise 1
Find the least common denominator of $\dfrac{7}{12}$ and $\dfrac{11}{20}$.	Find the least common denominator of $\dfrac{2}{15}$ and $\dfrac{17}{35}$.
SOLUTION: Begin by factoring each denominator. $$12 = 2 \cdot 2 \cdot 3$$ $$20 = 2 \cdot 2 \cdot 5$$ The least common denominator is equal to the product of the common factors $(2 \cdot 2)$ and the other factors $(3 \cdot 5)$. $$2 \cdot 2 \cdot 3 \cdot 5 = 60$$ The least common denominator is 60.	

Exercise 2

Find the least common denominator of $\frac{1}{24}$ and $\frac{25}{36}$.

Exercise 3

Find the least common denominator of $\frac{3}{14}$ and $\frac{13}{25}$.

Exercise 4

Find the least common denominator of $\frac{5}{18}$, $\frac{10}{27}$, and $\frac{16}{45}$.

Exercise 5

Find the least common denominator of $\frac{5}{12}$, $\frac{29}{30}$, and $\frac{49}{36}$.

1.1.5 Write equivalent fractions.

Two fractions are **equivalent fractions** if they are equal when simplified to lowest terms.

When adding or subtracting two fractions that have different denominators, you will need to first rewrite each fraction as an equivalent fraction with the least common denominator.

To rewrite a fraction as an equivalent fraction with a given denominator, determine what the original denominator must be multiplied by in order to produce the given denominator.

Then multiply the numerator by that same factor.

Multiplying both the numerator and denominator by the same number is equivalent to multiplying the fraction by 1, which does not change the value of the fraction.

Suppose you needed to write $\frac{3}{8}$ as an equivalent fraction whose denominator is 40.

First determine that you need to multiply 8 by 5 in order to equal 40.
Now multiply both the numerator and denominator by 5.

$$\frac{3}{8} = \frac{3 \cdot 5}{8 \cdot 5} = \frac{15}{40}$$

So, $\frac{15}{40}$ is the fraction that has a denominator of 40 and is equivalent to $\frac{3}{8}$.

Example	Exercise 1
Write $\frac{7}{12}$ as an equivalent fraction whose denominator is 48.	Write $\frac{3}{7}$ as an equivalent fraction whose denominator is 21.
SOLUTION: Since $48 \div 12 = 4$, multiply the numerator and denominator of $\frac{7}{12}$ by 4. $$\frac{7}{12} = \frac{7 \cdot 4}{12 \cdot 4}$$ $$= \frac{28}{48}$$	

Exercise 2

Write $\frac{5}{8}$ as an equivalent fraction whose denominator is 96.

Exercise 3

Write $\frac{2}{3}$ as an equivalent fraction whose denominator is 15.

Exercise 4

Write $\frac{6}{7}$ as an equivalent fraction whose denominator is 91.

Exercise 5

Write $\frac{14}{25}$ as an equivalent fraction whose denominator is 725.

1.1.6   Compare fractions.

In order to compare two fractions and determine which fraction is greater than or less than the other, they must have the same denominator.

If the two fractions have the same denominator, then the fraction with the lesser numerator is less than the other fraction.

Comparing Fractions
- Find the least common denominator of the fractions.

- Rewrite each fraction as an equivalent fraction whose denominator is the least common denominator.

- Compare the numerators.

Suppose you wanted to compare the fractions $\dfrac{5}{8}$ and $\dfrac{7}{10}$.

The least common denominator is 40. Rewrite $\dfrac{5}{8}$ as $\dfrac{25}{40}$ and rewrite $\dfrac{7}{10}$ as $\dfrac{28}{40}$.

Since $25 < 40$, then $\dfrac{25}{40} < \dfrac{28}{40}$. Thus $\dfrac{5}{8} < \dfrac{7}{10}$.

Example	**Exercise 1**
Write the correct sign (>, =, or <) between the two fractions: $\dfrac{5}{8}$ _____ $\dfrac{7}{12}$.	Write the correct sign (>, =, or <) between the two fractions: $\dfrac{4}{9}$ _____ $\dfrac{11}{18}$.
SOLUTION: The common denominator for these two fractions is 24. Rewrite each fraction as an equivalent fraction whose denominator is 24, then compare the fractions. $\dfrac{5}{8} = \dfrac{15}{24}$ $\qquad$ $\dfrac{7}{12} = \dfrac{14}{24}$ Since $\dfrac{15}{24} > \dfrac{14}{24}$, that means that $\dfrac{5}{8} > \dfrac{7}{12}$.	

Exercise 2

Write the correct sign (>, =, or <) between the two fractions: $\dfrac{41}{72}$ —— $\dfrac{13}{24}$.

Exercise 3

Write the correct sign (>, =, or <) between the two fractions: $\dfrac{19}{30}$ —— $\dfrac{17}{27}$.

Exercise 4

Write the correct sign (>, =, or <) between the two fractions: $\dfrac{48}{60}$ —— $\dfrac{68}{85}$.

Exercise 5

Write the correct sign (>, =, or <) between the two fractions: $\dfrac{71}{54}$ —— $\dfrac{47}{48}$.

1.1.7 Add or subtract unlike fractions.

When adding or subtracting two fractions that do not have the same denominator, we first find a common denominator by finding the least common denominator of the two denominators.

Then convert each fraction to an equivalent fraction whose denominator is that common denominator.

Once we rewrite the two fractions so they have the same denominator, we can add (or subtract) following the procedure for working with fractions that have the same denominator.

Adding or Subtracting Fractions with Unlike Denominators
- Find the least common denominator of the fractions.

- Rewrite each fraction as an equivalent fraction whose denominator is the least common denominator of the original denominators.

- Add or subtract the numerators, placing the result over the common denominator.

- Simplify to lowest terms.

$$\frac{5}{8} + \frac{7}{10} = \frac{25}{40} + \frac{28}{40}$$
$$= \frac{53}{40}$$

Example	Exercise 1
Simplify $\dfrac{11}{24} + \dfrac{5}{8}$.	Simplify $\dfrac{1}{2} - \dfrac{1}{3}$.
SOLUTION: The common denominator for these two fractions is 24. Rewrite each fraction as an equivalent fraction with the common denominator. Add the numerators and place the sum over the common denominator. Simplify the result to lowest terms. $\dfrac{11}{24} + \dfrac{5}{8} = \dfrac{11}{24} + \dfrac{15}{24}$ $= \dfrac{26}{24}$ $= \dfrac{13}{12}$	

Exercise 2

Simplify $\dfrac{2}{5} + \dfrac{3}{4}$.

Exercise 3

Simplify $\dfrac{2}{3} - \dfrac{7}{15}$.

Exercise 4

Simplify $\dfrac{7}{12} - \dfrac{5}{9}$.

Exercise 5

Simplify $\dfrac{22}{9} + \dfrac{7}{6}$.

1.1.8 Multiply or divide fractions.

To multiply fractions, multiply the numerators together and multiply the denominators together.

When multiplying fractions, you may simplify any fraction to lowest terms. You may also divide out a factor that is common to a numerator and a different denominator.

After all common factors have been divided out, multiply the remaining numerators and denominators.

For example, when multiplying $\dfrac{3}{5} \cdot \dfrac{7}{9}$ you can divide a common factor of 3 out of the first numerator and the second denominator.

$$\frac{3}{5} \cdot \frac{7}{9} = \frac{\overset{1}{\cancel{3}}}{5} \cdot \frac{7}{\underset{3}{\cancel{9}}} = \frac{7}{15}$$

To divide by a fraction, multiply by its reciprocal.

To find the reciprocal of a fraction, invert its numerator and denominator. The reciprocal of $\dfrac{4}{5}$ is $\dfrac{5}{4}$.

$$\frac{2}{3} \div \frac{4}{5} = \frac{2}{3} \cdot \frac{5}{4} = \frac{\overset{1}{\cancel{2}}}{3} \cdot \frac{5}{\underset{2}{\cancel{4}}} = \frac{5}{6}$$

Example	**Exercise 1**
Simplify $\dfrac{20}{21} \cdot \dfrac{49}{90}$.	Simplify $\dfrac{5}{8} \cdot \dfrac{6}{25}$.
SOLUTION: Begin by dividing out factors that are common to a numerator and a denominator. Once all common factors have been divided out, multiply the remaining factors in the numerators and the remaining factors in the denominators. $\dfrac{20}{21} \cdot \dfrac{49}{90} = \dfrac{\overset{2}{\cancel{20}}}{21} \cdot \dfrac{49}{\underset{9}{\cancel{90}}}$ $= \dfrac{2}{\underset{3}{\cancel{21}}} \cdot \dfrac{\overset{7}{\cancel{49}}}{9}$ $= \dfrac{2}{3} \cdot \dfrac{7}{9}$ $= \dfrac{14}{27}$	

Exercise 2

Simplify $\dfrac{12}{25} \div \dfrac{8}{45}$.

Exercise 3

Simplify $\dfrac{11}{28} \div \dfrac{33}{147}$.

Exercise 4

Simplify $\dfrac{28}{5} \div 7$.

Exercise 5

Simplify $\dfrac{11}{42} \cdot \dfrac{35}{99}$.

1.1.9 Evaluate exponential expressions with fractional bases.

If n is a positive integer, the expression b^n is equivalent to the product where the base b is listed as a factor n times.

This remains true if the base of the exponential expression is a fraction.

For example, the expression $\left(\dfrac{5}{6}\right)^3$ can be rewritten as the product $\dfrac{5}{6} \cdot \dfrac{5}{6} \cdot \dfrac{5}{6}$.

Once the expression has been rewritten as a product, multiply the numerators and multiply the denominators.

$$\left(\frac{5}{6}\right)^3 = \frac{5}{6} \cdot \frac{5}{6} \cdot \frac{5}{6} = \frac{125}{216}$$

Notice that the expression $\left(\dfrac{5}{6}\right)^3$ is equivalent to the expression $\dfrac{5^3}{6^3}$.

This gives an alternative strategy for raising a fraction to a power: raise both the numerator and the denominator to that power.

Example	Exercise 1
Simplify $\left(\dfrac{2}{5}\right)^4$.	Simplify $\left(\dfrac{1}{8}\right)^3$.
SOLUTION: Rewrite the expression as a product with $\dfrac{2}{5}$ as a factor 4 times. $\left(\dfrac{2}{5}\right)^4 = \dfrac{2}{5} \cdot \dfrac{2}{5} \cdot \dfrac{2}{5} \cdot \dfrac{2}{5}$ $= \dfrac{16}{625}$ Another approach would be to raise both the numerator and denominator to the fourth power, then simplify. $\left(\dfrac{2}{5}\right)^4 = \dfrac{2^4}{5^4}$ $= \dfrac{16}{625}$	

Exercise 2

Simplify $\left(\dfrac{3}{2}\right)^2$.

Exercise 3

Simplify $\left(\dfrac{4}{7}\right)^5$.

Exercise 4

Simplify $\left(\dfrac{14}{15}\right)^2$.

Exercise 5

Simplify $\left(\dfrac{1}{10}\right)^7$.

1.1.10 Use the order of operations on fractions.

The **order of operations** agreement is a standard order in which arithmetic operations are performed, ensuring a single correct result.

Order of Operations
1. **Remove grouping symbols.**
   Begin by simplifying all expressions within parentheses, brackets, and absolute value bars. Also, perform any operations in the numerator and denominator of a fraction. This is done by following Steps 2 – 4, presented next.
2. **Perform any operations involving exponents.**
   After all operations have been performed inside grouping symbols, simplify any exponential expressions.
3. **Multiply and divide.**
   These two operations have equal priority. Find products or quotients in the order that they appear from left to right.
4. **Add and subtract.**
   At this point, the only remaining operations should be addition and subtraction. Again, these operations are of equal priority, and they are performed in the order they appear from left to right.

Once you determine which operation to perform, follow the rules for fraction arithmetic with that type of operation.

<table>
<tr><td>

Example

Simplify: $\dfrac{2}{5} + \dfrac{15}{7} \cdot \dfrac{21}{20}$

SOLUTION:

Begin by multiplying $\dfrac{15}{7} \cdot \dfrac{21}{20}$.

Then add that sum to $\dfrac{2}{5}$.

$$\frac{2}{5} + \frac{15}{7} \cdot \frac{21}{20} = \frac{2}{5} + \frac{\overset{3}{\cancel{15}}}{\underset{1}{\cancel{7}}} \cdot \frac{\overset{3}{\cancel{21}}}{\underset{4}{\cancel{20}}}$$

$$= \frac{2}{5} + \frac{9}{4}$$

$$= \frac{8}{20} + \frac{45}{20}$$

$$= \frac{53}{20}$$

</td><td>

Exercise 1

Simplify: $\dfrac{1}{2} + \dfrac{1}{2} \cdot \dfrac{4}{7}$

</td></tr>
</table>

Exercise 2

Simplify: $\dfrac{1}{9}\left(\dfrac{19}{28}-\dfrac{1}{4}\right)$

Exercise 3

Simplify: $\dfrac{5}{6}+\left(\dfrac{3}{2}\right)^2$

Exercise 4

Simplify: $\dfrac{8}{25}\div\left(\dfrac{4}{15}-\dfrac{1}{3}+\dfrac{3}{5}\right)\cdot\dfrac{7}{18}$

Exercise 5

Simplify: $\dfrac{3}{8}+\dfrac{1}{8}\div\left(\dfrac{5}{4}\right)^2-\dfrac{7}{16}$

1.1.11 Solve applications involving fractions.

To solve an application problem, you must decide which operation or operations are to be performed, and in what order.

Key Ideas
- Addition: Problems involving a total of two or more quantities

- Subtraction: Problems involving the difference between two quantities, or how much remains after one quantity is taken away from another quantity

- Multiplication: Problems involving one quantity being used multiple times

- Division: Problems involving breaking up a quantity into equally sized quantities

Once you have determined which operations need to be performed, keep in mind how the procedures for addition, subtraction, multiplication, and division of fractions are performed.

If you struggle to determine which operations to perform, consider replacing the fractions with appropriate integers and think about this new problem. Then you can return to the problem involving fractions and use the same operations.

Example	Exercise 1
Three recipes call for $\frac{1}{2}$ cup, $\frac{3}{4}$ cup, and $\frac{1}{3}$ cup of flour, respectively. In total, how much flour is needed to make these three recipes? SOLUTION: To find the total amount of flour, add the three fractions together. $$\frac{1}{2}+\frac{3}{4}+\frac{1}{3}=\frac{6}{12}+\frac{9}{12}+\frac{4}{12}$$ $$=\frac{19}{12}$$ Rewrite the sum $\frac{19}{12}$ as the mixed number $1\frac{7}{12}$. In total, $1\frac{7}{12}$ cups of flour are needed to make these three recipes.	A mother gave birth to twins. They weighed $5\frac{7}{8}$ pounds and $5\frac{1}{4}$ pounds. Find the total weight of the twins.

Exercise 2

A chemist had $\frac{29}{40}$ fluid ounces of a solution. If she used $\frac{3}{8}$ fluid ounces of the solution in an experiment, how much of the solution remains?

Exercise 3

Marcus had $7\frac{1}{4}$ cups of chicken stock before making a stew. If he now has $3\frac{3}{8}$ cups of stock left, how much did he put in the stew?

Exercise 4

A walkway is $\frac{15}{32}$ miles long. If the organizers of an event want to divide the walkway into 40 sections of equal length, how long will each section be?

Exercise 5

The directions on a bag of fertilizer recommend using $\frac{2}{3}$ pound of fertilized for every 1000 square feet of lawn. How much fertilizer should be used for an 9000-square foot lawn?

1.2.1   Identify place values of decimals.

Just as with whole numbers, each digit that appears after a decimal point has a place value associated with it.

- The first digit after the decimal point is in the tenths place.
- The second digit after the decimal point is in the hundredths place.
- The third digit after the decimal point is in the thousandths place.
- The pattern continues in the same way for successive digits.

Consider the number 0.123456789.
- The digit 1 is in the tenths place.
- The digit 2 is in the hundredths place.
- The digit 3 is in the thousandths place.
- The digit 4 is in the ten-thousandths place.
- The digit 5 is in the hundred-thousandths place.
- The digit 6 is in the millionths place.
- The digit 7 is in the ten-millionths place.
- The digit 8 is in the hundred-millionths place.
- The digit 9 is in the billionths place.

Example	**Exercise 1**
What is the decimal place of the digit 3 in the number 5.1239? SOLUTION: The digit 3 is located three places to the right of the decimal point, which is the thousandths place.	What is the decimal place of the digit 3 in the number 0.304?

Exercise 2
What is the decimal place of the digit 3 in the number 17.0003?

Exercise 3
What is the decimal place of the digit 3 in the number 6.23?

Exercise 4
What is the decimal place of the digit 3 in the number 45.891234?

Exercise 5
What is the decimal place of the digit 3 in the number 0.876543?

1.2.2 Compare decimals.

Comparing decimal numbers is similar to comparing whole numbers.

Compare the numbers by their place values from left to right.

The first place value where the two numbers differ can be used to determine which number is greater than the other.

For example, suppose you had to put the numbers 17.2389, 17.2265, and 17.238 in ascending order.

All three numbers agree in the tens and ones place.

To the right of the decimal point, all three numbers have the digit 2 in the tenths place.

There is a difference in the hundredths place. 17.2269 has a 2 in the hundredths place, while the other two numbers have a 3 in the hundredths place. Since 2 is less than 3, 17.2269 is the smallest number in the list.

The other two numbers agree in the thousandths place, but there is a difference in the ten-thousandths place. Although we do not see a digit in the ten-thousandths place of 17.238, it is a 0. 17.238 is the same as 17.2380. Since 0 is less than 9, we know that 17.238 is less than 17.2389.

So, rewriting the list in ascending order yields 17.2269, 17.238, 17.2389.

Example	Exercise 1
Write the correct sign (> or <) between the two numbers: 0.05 _____ 0.0063	Write the correct sign (> or <) between the two numbers: 0.05 _____ 0.053
SOLUTION: The first decimal place where the numbers differ is the hundredths place. Since $5 > 0$, $0.05 > 0.0063$.	

Exercise 2 Write the correct sign (> or <) between the two numbers: 0.00001 _____ 0.000001	**Exercise 3** Write the correct sign (> or <) between the two numbers: 0.1009 _____ 0.01
Exercise 4 Write the correct sign (> or <) between the two numbers: 0.0432 _____ 0.043201	**Exercise 5** Write the correct sign (> or <) between the two numbers: 0.0295 _____ 0.0259

1.2.3 Add or subtract decimals.

Adding or subtracting decimal numbers is similar to adding or subtracting whole numbers.

Align the two numbers vertically by their place values. The decimal point in the first number should be directly over the decimal point in the second number.

If one number has more decimal places than the other, you can add trailing 0's to the end of one number so the number of decimal places match.

If the sum in one decimal place is 10 or greater, carry to the next place value as you would when adding whole numbers.

Here is how you would add $3.578 + 4.6$:

$$
\begin{array}{r}
\overset{1}{3.578} \\
+\quad 4.600 \\
\hline
8.178
\end{array}
$$

You can also borrow while subtracting, just as you would when subtracting whole numbers.

Example	**Exercise 1**
Simplify: $4.29 + 8.62$	Simplify: $9.4 - 3.7$
SOLUTION: Align the numbers vertically by place values. $\begin{array}{r}\overset{1}{4.29} \\ +\quad 8.62 \\ \hline 12.91\end{array}$	

Exercise 2
Simplify: $15.2 - 8.75$

Exercise 3
Simplify: $9.6795 + 13.3208$

Exercise 4
Simplify: $7.3 - 2.95 + 18.657$

Exercise 5
Simplify: $-9.34 + 12.543 - 31.8$

1.2.4 Multiply or divide decimals.

Multiplication
To multiply two decimal numbers, multiply them as you would multiply whole numbers.

The number of decimal places in the product is equal to the total number of decimal places in the two factors.

For example, in order to multiply $5.13 \cdot 2.2$, begin by multiplying $513 \cdot 22 = 11286$. The first factor has 2 decimal places and the second factor has one decimal place, so the product will have 3 decimal places. Rewrite the product so that it has 3 digits after the decimal point: $5.13 \cdot 2.2 = 11.286$.

Division
To divide by a decimal number, rewrite the problem so that the divisor is a whole number.
Move the decimal point in the divisor to the right so that it becomes a whole number.
Move the decimal point in the dividend to the right by the same number of places.

For example, rewrite the problem $8.13 \div 0.6$ as $81.3 \div 6$.

After setting up the problem as a long division problem, the decimal point in the quotient will be directly above the decimal point in the dividend.

If there is a remainder, continue to add 0's to the end of the dividend until there is no remainder or the pattern of a repeating decimal appears.

Example	**Exercise 1**
Simplify: $3.2 \cdot 4.8$	Simplify: $13.568 \div 0.4$
SOLUTION: The first factor has one decimal place, and so does the second. The product will have two decimal places. $\begin{array}{r} 3.2 \\ \times 4.8 \\ \hline 256 \\ 1280 \\ \hline 15.36 \end{array}$	

Exercise 2
Simplify: $2.4 \cdot 3.65$

Exercise 3
Simplify: $69.54 \div 6$

Exercise 4
Simplify: $7.2(-2.1)(-3.06)$

Exercise 5
Simplify: $-7.156 \div 0.25$

1.2.5 Evaluate exponential expressions with decimal bases.

If n is a positive integer, the expression b^n is equivalent to the product where the base b is listed as a factor n times.

This remains true if the base of the exponential expression is a decimal number.

For example, the expression 0.4^3 can be rewritten as the product
$$0.4 \cdot 0.4 \cdot 0.4$$

Once the expression has been rewritten as a product, multiply the decimal numbers.

$$
\begin{aligned}
0.4^3 &= 0.4 \cdot 0.4 \cdot 0.4 \\
&= 0.16 \cdot 0.4 \\
&= 0.064
\end{aligned}
$$

Keep in mind that the number of decimal places in the product will equal the total number of decimal places in all of the factors.

Example	Exercise 1
Simplify 0.2^5.	Simplify 0.5^4.
SOLUTION: Use 0.2 as a factor five times. Since each factor has one digit after the decimal point, the product must have five digits after the decimal point. $$\begin{aligned}0.2^5 &= 0.2 \cdot 0.2 \cdot 0.2 \cdot 0.2 \cdot 0.2 \\ &= 0.00032\end{aligned}$$	

Exercise 2

Simplify 3.6^2.

Exercise 3

Simplify -1.6^2.

Exercise 4

Simplify $(-1.6)^2$.

Exercise 5

Simplify $2.2^2 \cdot 0.1^3$.

1.2.6 Use the order of operations on decimals.

The **order of operations** agreement is a standard order in which arithmetic operations are performed, ensuring a single correct result.

Order of Operations
1. **Remove grouping symbols.**
   Begin by simplifying all expressions within parentheses, brackets, and absolute value bars. Also, perform any operations in the numerator and denominator of a fraction. This is done by following Steps 2 – 4, presented next.
2. **Perform any operations involving exponents.**
   After all operations have been performed inside grouping symbols, simplify any exponential expressions.
3. **Multiply and divide.**
   These two operations have equal priority. Find products or quotients in the order that they appear from left to right.
4. **Add and subtract.**
   At this point, the only remaining operations should be addition and subtraction. Again, these operations are of equal priority, and they are performed in the order they appear from left to right.

Once you determine which operation to perform, follow the rules for arithmetic with decimal numbers for that type of operation.

Example	**Exercise 1**
Simplify: $2 \cdot 3.2 + 5.8 \cdot 2.5$	Simplify: $2.6 + 2.4 \cdot 5.3$
SOLUTION: Since there are no grouping symbols or exponents, begin by finding the two products. $$2 \cdot 3.2 + 5.8 \cdot 2.5 = 6.4 + 14.5$$ $$= 20.9$$	

Exercise 2

Simplify: $6.5^2 - 3.9^2$

Exercise 3

Simplify: $6(16.8 - 4.4 \cdot 7.25)$

Exercise 4

Simplify: $14.4 \div 3.2 \cdot 21.3$

Exercise 5

Simplify: $7.5(2.8 + 3.6 \cdot 0.2 - 5.5^2)$

1.2.7   Round decimals.

Rounding a decimal number to a given place value is similar to rounding whole numbers.

Rounding Decimals
1. Identify the digit in the place the number is to be rounded.

2. Examine the digit immediately to its right.

3. If the digit to the right is 5 or greater, round the number up.

    If the digit to the right is 4 or less, round the number down.

When rounding a decimal number to a certain place, leave off all trailing numbers after that specified place.

Suppose you were asked to round 8.2963 to the nearest hundredth.
The digit in the hundredths place is a 9, and the digit to the right of that is a 6.
Since the digit to the right is 5 or greater, round the number up by adding 1 to the 9 in the hundredths place. The rounded number should end in the hundredths place: 8.30.

Example Round 17.47 to the nearest tenth. SOLUTION: The digit in the tenth place is a 4. The digit to the right of that is a 7. Since the digit in the hundredths place is 5 or greater, round up to 17.5.	**Exercise 1** Round 6.0348 to the nearest hundredth.

Exercise 2 Round 23.96 to the nearest tenth.	**Exercise 3** Round 46.91827364 to the nearest ten-thousandth.
Exercise 4 Round 46.91827364 to the nearest ten-millionth.	**Exercise 5** Round 7.003482 to the nearest hundredth.

1.2.8   Solve problems involving estimation with decimals.

Being able to estimate calculations is an important skill.

It allows you to determine whether your calculation is accurate.

One estimation strategy is to round each number to the nearest whole number, and then perform the calculation using the whole numbers.

For example, suppose you wanted to estimate the product $15.75 \cdot 4.3$.

First round 15.75 to the nearest whole number, which is 16.
Next round 4.3 to the nearest whole number, which is 4.

To estimate the product, multiply $16 \cdot 4$, which is 64. So, 64 is an estimate of the product $15.75 \cdot 4.3$. (The actual product is 67.725. The difference between the estimate and the actual product is due to rounding each of the factors first.)

Example	Exercise 1
Round each value to the nearest whole number before calculating to estimate the result of: $$7.8 \cdot 3.2 - 15.9$$ SOLUTION: Begin by rounding each number to the nearest whole number. $7.8 \approx 8$ $3.2 \approx 3$ $15.9 \approx 16$ Replace each value and simplify. $$7.8 \cdot 3.2 - 15.9 \approx 8 \cdot 3 - 16$$ $$= 24 - 16$$ $$= 8$$	Round each value to the nearest whole number before calculating to estimate the result of: $$23.762 \div 7.95$$

Exercise 2
Round each value to the nearest whole number before calculating to estimate the result of:
$$32.6 \cdot 12.4 - 9.27^2$$

Exercise 3
Round each value to the nearest whole number before calculating to estimate the result of:
$$21.9 \cdot 9.8 - 5.4785 \cdot 19.875$$

Exercise 4
A shopper buys 12 copies of a book that sells for $8.88. Estimate the total cost.

Exercise 5
A shopper bought 20 ornaments for $67.97. Estimate the cost of each ornament.

1.2.9 Convert decimals to fractions.

To convert a decimal number to a fraction, begin by determining the place value of the last digit. That place value will tell you the denominator of the fraction.

For example, if the last digit is in the thousandths place then the denominator should be 1000. The denominator of a fraction that is 1000 is read as thousandths.

Write the decimal number, without the decimal point, in the numerator of the fraction.

Simplify the resulting fraction to lowest terms. Since the denominator will always start as a power of 10, the only possibilities for common factors between the numerator and denominator are 2 and 5.

For example, suppose you had to rewrite 0.864 as a fraction is lowest terms.
- The last digit, 4, is in the thousandths place.
- The denominator of the fraction will be 1000.
- The numerator of the fraction is 864.
- Simplify to lowest terms, by continuing to divide out common factors of 2.

$$\frac{864}{1000} = \frac{432}{500} = \frac{216}{250} = \frac{108}{125}$$

So, $0.864 = \frac{108}{125}$.

Example	**Exercise 1**
Rewrite 0.75 as a fraction in lowest terms.	Rewrite 0.4 as a fraction in lowest terms.
SOLUTION:	
The last digit, 5, is in the hundredths place. The denominator for the initial fraction will be 100.	
Begin by writing the fraction $\frac{75}{100}$ and then simplify the fraction to lowest terms.	
$$\frac{75}{100} = \frac{3}{4}$$	

Exercise 2
Rewrite 0.68 as a fraction in lowest terms.

Exercise 3
Rewrite 0.325 as a fraction in lowest terms.

Exercise 4
Rewrite 0.032 as a fraction in lowest terms.

Exercise 5
Rewrite 0.2004 as a fraction in lowest terms.

1.2.10 Convert fractions to decimals.

To rewrite a fraction as a decimal number, divide its numerator by its denominator.

For example, suppose you needed to rewrite $\dfrac{9}{16}$ as a decimal number.

Divide the numerator by the denominator: $9 \div 16 = 0.5625$.

So, $\dfrac{9}{16} = 0.5625$.

Occasionally, when you divide the numerator by the denominator, the quotient will not be a decimal that terminates. Instead, you may see a repeating decimal.

Write the decimal number with a bar over the repeating portion.

Consider the fraction $\dfrac{2}{11}$. When you divide, $2 \div 11 = 0.181818...$.

So, $\dfrac{2}{11} = 0.\overline{18}$.

Example	**Exercise 1**
Rewrite $\dfrac{5}{8}$ as a decimal number.	Rewrite $\dfrac{7}{10}$ as a decimal number.
SOLUTION: To rewrite $\dfrac{5}{8}$ as a decimal number, divide 5 by 8. $$5 \div 8 = 0.625$$ So, $\dfrac{5}{8} = 0.625$.	

Exercise 2

Rewrite $\dfrac{29}{1000}$ as a decimal number.

Exercise 3

Rewrite $\dfrac{17}{25}$ as a decimal number.

Exercise 4

Rewrite $\dfrac{63}{4}$ as a decimal number.

Exercise 5

Rewrite $7\dfrac{3}{20}$ as a decimal number.

1.2.11 Solve applications involving decimals.

To solve an application problem, you must decide which operation or operations are to be performed, and in what order.

Key Ideas
- Addition: Problems involving a total of two or more quantities

- Subtraction: Problems involving the difference between two quantities, or how much remains after one quantity is taken away from another quantity

- Multiplication: Problems involving one quantity being used multiple times

- Division: Problems involving breaking up a quantity into equally sized quantities

Once you have determined which operations need to be performed, keep in mind how the procedures for addition, subtraction, multiplication, and division of decimal numbers are performed.

If you struggle to determine which operations to perform, consider replacing the decimals with appropriate integers and think about this new problem. Then you can return to the problem involving decimals and use the same operations.

Example	**Exercise 1**
A shopper buys 12 copies of a book that sells for $8.88. Find the total cost. SOLUTION: To find the cost of 12 copies of the book, multiply the cost of one book ($8.88) by 12. $$\$8.88 \cdot 12 = \$106.56$$ The total cost is $106.56.	A shopper bought 20 ornaments for $67.80. Find the cost of each ornament.

Exercise 2

Cardi has a fever and her temperature is 101.4°F. How far above the normal body temperature (98.6°F) is her temperature?

Exercise 3

Paul spent the following amounts on gifts: $32.95, $27.99, $42.50, and $199. How much did Paul spend in total on gifts?

Exercise 4

At the beginning of the week the price of a stock was $167.29. Over the five days of the week it went up by $3.67, down by $8.05, down by $0.16, up by $5.99, and down by $3.76. What was the price of the stock at the end of the week?

Exercise 5

Chris heads to the store with a $20 bill to buy as many packages of hot dogs as possible. If each package costs $2.65, how much change from the $20 bill does Chris come home with?

1.3.1   Interpret the meaning of percent.

Percents (%) are used to represent numbers as parts of 100.

One percent, which can be written as 1%, is equivalent to 1 part of 100, or $\dfrac{1}{100}$, or 0.01.

The number 27% represents 27 parts of 100, or $\dfrac{27}{100}$.

Percents can be used to compare parts of two groups of different sizes, such as the percent of students at a small college (enrollment: 4000) who are first-generation college students compared to the percent of students at a large university (enrollment: 50,000) who are first-generation college students.

This comparison of percents is a better comparison than simply comparing the total number of first-generation students at each school because of the different size of each school.

Example	**Exercise 1**
How many parts of 100 is 24%? SOLUTION: One percent is 1 part of 100, so 24% is 24 parts of 100.	How many parts of 100 is 8%?

Exercise 2
Is 53% less than or greater than one-half?

Exercise 3
Is 70% less than or greater than three-quarters?

Exercise 4
Is 0.03 less than 5%?

Exercise 5
Is 0.013 less than 1%?

1.3.2 Write fractions as percents.

To rewrite a fraction as a percent, multiply it by 100%.

Because 100% is equivalent to 1, multiplying by 100% does not change the value of the fraction but allows us to rewrite it in a different form.

For example, suppose you wanted to rewrite the fraction $\frac{9}{20}$ as a percent. Start by multiplying it by 100%.

$$\frac{9}{20} \cdot 100\% = \frac{9}{\underset{1}{\cancel{20}}} \cdot \overset{5}{\cancel{100}}\% = 45\%$$

(You may find it makes more sense to rewrite 100% as the fraction $\frac{100\%}{1}$. This may make it easier for you to see that you can divide out common factors from the denominator and 100.)

If there remains a factor in the denominator after multiplying by 100%, write the percent using a mixed number.

Here is how to rewrite the fraction $\frac{4}{9}$ as a percent using a mixed number.

$$\frac{4}{11} \cdot 100\% = \frac{400}{11}\% = 36\frac{4}{11}\%$$

Example	**Exercise 1**
Rewrite $\frac{4}{5}$ as a percent.	Rewrite $\frac{1}{4}$ as a percent.
SOLUTION: Multiply the fraction by 100%. $$\frac{4}{5} \cdot 100\% = \frac{4}{\underset{1}{\cancel{5}}} \cdot \overset{20}{\cancel{100}}\%$$ $$= 80\%$$	

Exercise 2

Rewrite $\dfrac{3}{8}$ as a percent.

Exercise 3

Rewrite $\dfrac{7}{16}$ as a percent.

Exercise 4

Rewrite $\dfrac{5}{12}$ as a percent.

Exercise 5

Rewrite $\dfrac{19}{4}$ as a percent.

1.3.3   Write percents as fractions in simplest form.

To rewrite a percent as a fraction, divide it by 100 and omit the percent sign.

Alternatively, you can omit the percent sign and multiply by the fraction $\dfrac{1}{100}$.

For example, suppose you needed to rewrite 48% as a fraction.

Start with the fraction $\dfrac{48}{100}$, omitting the percent sign, and then simplify to lowest terms.

$$48\% = \frac{48}{100} = \frac{\overset{12}{\cancel{48}}}{\underset{25}{\cancel{100}}} = \frac{12}{25}$$

The prime factorization of 100 is $2 \cdot 2 \cdot 5 \cdot 5$, so when simplifying to lowest terms you should look for factors of 2 or 5 in the numerator that can be divided out of the numerator and denominator.

If the percent is 100% or larger, the resulting fraction will be a whole number or improper fraction.

If the percent contains a mixed number, first convert the mixed number to an improper fraction, omit the percent sign, and multiply by $\dfrac{1}{100}$.

$$6\frac{1}{4}\% = \frac{25}{4}\% = \frac{25}{4} \cdot \frac{1}{100} = \frac{\overset{1}{\cancel{25}}}{4} \cdot \frac{1}{\underset{4}{\cancel{100}}} = \frac{1}{16}$$

Example	Exercise 1
Rewrite 64% as a fraction in simplest form.	Rewrite 7% as a fraction in simplest form.
SOLUTION: Rewrite 64% as $\dfrac{64}{100}$ and simplify to lowest terms. $$64\% = \frac{64}{100}$$ $$= \frac{16}{25}$$	

Exercise 2
Rewrite 5% as a fraction in simplest form.

Exercise 3
Rewrite 520% as a fraction in simplest form.

Exercise 4
Rewrite 60% as a fraction in simplest form.

Exercise 5
Rewrite $11\frac{1}{9}$% as a fraction in simplest form.

1.3.4 Write decimals as percents.

To rewrite a decimal as a percent, multiply by 100%.

Keep in mind that multiplying a decimal by 100 is equivalent to moving the decimal point 2 places to the right.

Consider the following examples:

$$0.34 = 0.34 \cdot 100\% = 34\%$$

$$0.1 = 0.1 \cdot 100\% = 10\%$$

$$0.002 = 0.002 \cdot 100\% = 0.2\%$$

$$4 = 4 \cdot 100\% = 400\%$$

Any number greater than or equal to 1 will be equivalent to a percent that is greater than or equal to 100%.

Example	**Exercise 1**
Rewrite 0.6 as a percent.	Rewrite 0.87 as a percent.
SOLUTION: Multiply by 100% and simplify the product. $$0.6 = 0.6 \cdot 100\%$$ $$= 60\%$$	

Exercise 2
Rewrite 0.03 as a percent.

Exercise 3
Rewrite 0.15 as a percent.

Exercise 4
Rewrite 2.4 as a percent.

Exercise 5
Rewrite 0.155 as a percent.

1.3.5 Write percents as decimals.

To write a percent as a decimal, omit the percent time and divide by 100.

Keep in mind that dividing a decimal number or whole number by 100 is equivalent to moving the decimal point 2 places to the left.

Consider the following examples:

$$28\% = 28 \div 100 = 0.28$$

$$3\% = 3 \div 100 = 0.03$$

$$17.5\% = 17.5 \div 100 = 0.175$$

$$800\% = 800 \div 100 = 8$$

Example	**Exercise 1**
Write 71% as a decimal.	Write 16% as a decimal.
SOLUTION: Divide by 100 and drop the percent sign. $$71\% = 71 \div 100$$ $$= 0.71$$	

Exercise 2
Write 9% as a decimal.

Exercise 3
Write 3.4% as a decimal.

Exercise 4
Write 400% as a decimal.

Exercise 5
Write 0.5% as a decimal.

1.3.6 Convert among fractions, decimals, and percents.

Converting a Fraction to a Decimal Number
- Divide the numerator by the denominator.

Converting a Decimal Number to a Fraction
- Use the place value of the decimal number to determine the denominator.
- Write the decimal number in the numerator without its decimal point.
- Simplify the fraction to lowest terms.

Converting a Fraction to a Percent
- Multiply by 100%.

Converting a Percent to a Fraction
- Divide by 100 and omit the percent sign.

Converting a Decimal Number to a Percent
- Multiply by 100%.

Converting a Percent to a Decimal Number
- Divide by 100 and omit the percent sign.

Example
Complete the table.

Fraction	Decimal	Percent
	0.2	

SOLUTION:
To rewrite 0.2 as a fraction, begin with a fraction whose numerator is 2 and whose denominator is 10.
Simplify to lowest terms.

$$0.2 = \frac{2}{10}$$
$$= \frac{1}{5}$$

To rewrite 0.2 as a percent, multiply by 100%.

$$0.2 = 0.2 \cdot 100\%$$
$$= 20\%$$

Exercise 1
Complete the table.

Fraction	Decimal	Percent
$\frac{11}{20}$		

Exercise 2
Complete the table.

Fraction	Decimal	Percent
		24%

Exercise 3
Complete the table.

Fraction	Decimal	Percent
		25%

Exercise 4
Complete the table.

Fraction	Decimal	Percent
$\frac{7}{8}$		

Exercise 5
Complete the table.

Fraction	Decimal	Percent
	0.88	

1.3.7 Perform calculations involving percents.

To find a certain percentage of a number called the base, multiply the percent by the base. Before multiplying, convert the percent to a decimal number.

For example, suppose you needed to find 20% of 325.
You can do this by multiplying 20% by 325.

$$20\% \cdot 325 = 0.2 \cdot 325 = 65$$

So, 20% of 325 is 65.

In some cases, it may be a good choice to convert the percent to a fraction before multiplying. For example, this is an effective strategy when the percent is in the form of a mixed number.

Suppose you needed to find $8\frac{3}{4}\%$ of 400.

First, rewrite $8\frac{3}{4}\%$ as a fraction by multiplying by $\frac{1}{100}$. Omit the percent sign.

$$8\frac{3}{4}\% = \frac{35}{4} \cdot \frac{1}{100} = \frac{\overset{7}{\cancel{35}}}{4} \cdot \frac{1}{\underset{20}{\cancel{100}}} = \frac{7}{80}$$

Now multiply the fraction by the base 400.

$$\frac{7}{80} \cdot 400 = \frac{7}{\underset{1}{\cancel{80}}} \cdot \overset{5}{\cancel{400}} = 35$$

So, $8\frac{3}{4}\%$ of 400 is 35.

Example	Exercise 1
Find 40% of 85.	Find 75% of 224.
SOLUTION:	
To find 40% of 85, rewrite 40% as a decimal and multiply by 85.	
$40\% = 0.4$	
$\qquad 0.4 \cdot 85 = 34$	
40% of 85 is 34.	

Exercise 2
Find 37.5% of 800.

Exercise 3
Find 6% of 17,550.

Exercise 4
Find $7\frac{1}{2}$% of 100,000.

Exercise 5
Find 0.3% of 120.

1.3.8 Compute percent change.

Percent increase is a measure of how much a quantity has increased from its original value.

The amount of increase is equal to the percent increase multiplied by the original value.

Amount of increase = Percent increase · Original value

Suppose a worked receives a pay increase from $16/hour to $20/hour. What is the percent increase in the hourly pay rate?
The amount of increase is $20 - $16 = $4. Let x represent the percent increase. The original value is $16.

$$4 = x \cdot 16$$
$$\frac{4}{16} = \frac{x \cdot 16}{16}$$
$$\frac{1}{4} = x$$

Now convert the fraction to a percent by multiplying by 100%: $\frac{1}{4} \cdot 100\% = \frac{1}{\cancel{4}} \cdot \cancel{100}^{25} \% = 25\%$.

The percent increase in the hourly pay rate is 25%.

Percent decrease is a similar measure of how much a quantity has decreased from its original value. A similar formula can be used for problems involving percent decrease.

Amount of decrease = Percent decrease · Original value

Example	Exercise 1
Marlana invested $3500 in the stock market. In one year, the investment had increased to $4095. Find the percent increase in Marlana's investment.	The price of a gallon of milk increased from $3.25 to $3.51. Find the percent increase in the price.
SOLUTION: Begin by computing the increase. $4095 - $3500 = $595	
Next, determine what percent of the original investment ($3500) is the increase ($595). Let p represent the percent increase. $$p \cdot 3500 = 595$$ $$p = \frac{595}{3500}$$ $$p = 0.17$$ $$p = 17\%$$ The percent increase in Marlana's investment is 17%.	

Exercise 2

A bookstore pays a wholesale price of $24. They mark up the book and sell it for $31.68. Find the percent increase applied to the cost of the book.

Exercise 3

A community college had 1600 parking spots, but a building project cut the number of spaces down to 1480 spaces. Find the percent decrease in the number of parking spots.

Exercise 4

A community college's annual budget dropped from $40 million to $37.2 million. Find the percent decrease in the community college's annual budget.

Exercise 5

A sweater's original price was $59.95. During a sale it was marked down to $47.96. Find the percent decrease in the price of the sweater.

1.3.9 Solve applications involving percents.

To solve applications involving percents, use the basic percent equation.

$$\text{Percent} \cdot \text{Base} = \text{Amount}$$

The percent, if known, should be written as a decimal number or as a fraction.

If the percent is unknown, convert it from a fraction or decimal number to a percent after solving the equation.

A community college has 4200 students. Forty percent of those students plan to transfer to a 4-year university. How many students plan to transfer to a 4-year university.

- The percent is 40%, which is 0.4 as a decimal number.
- The base is the group of 4200 students.
- The amount is unknown. Let x represent the number of students who plan transfer to a 4-year university.

$$\text{Percent} \cdot \text{Base} = \text{Amount}$$
$$0.4(4200) = x$$
$$1680 = x$$

1680 of the students plan to transfer to a 4-year university.

Example	Exercise 1
Thirty percent of the candies in a bowl are red. If there are 200 candies in the bowl, how many are red? SOLUTION: Let x represent the number of red candies. Since 30% of the candies are red, find 30% of 200 to find the number of red candies. $$0.3 \cdot 200 = x$$ $$60 = x$$ There are 60 red candies in the bowl.	A doctor helped 320 women give birth last year. Sixteen of these women had twins. What percent of her patients had twins?

Exercise 2

Forty percent of the registered voters in a certain precinct are independent. If 410 of the registered voters in that precinct are independent, how many registered voters are in the precinct?

Exercise 3

An online retailer adds a 5% charge to all items for shipping and handling. If a shipping-and-handling fee of $7.25 is added to the cost of a coffee maker, what is the original price of the coffee maker?

Exercise 4

A certain brand of rum is 40% alcohol. How many milliliters of alcohol are in a 750-milliliter bottle of that brand of rum?

Exercise 5

A bank pays 0.64% annual interest on certificates of deposit (CDs). Find the amount of interest earned in one year on a $800 CD.

1.4.1 Write ratios in different notations, including fractions.

A **ratio** is a comparison of two quantities with the same unit, and can be represented in multiple ways.

Suppose a class has 40 students in it. Four of the students are left-handed and 36 are right-handed. One way to express the ratio of left-handed students to right-handed students is in a sentence using the word *to*:

4 students to 36 students

or simply

4 to 36

The ordering of the quantities is important. The first quantity mentioned will always be written first.

The same ratio can also be expressed using a colon (:) between the two quantities.

4:36

The ratio can also be expressed as a fraction, with the first quantity in the numerator and the second quantity in the denominator.

$$\frac{4}{36}$$

Simplifying ratios is similar to simplifying a fraction to lowest terms, and will be covered on another worksheet.

Example	**Exercise 1**
One out of every 9 people are left-handed. Write the ratio of left-handed people to all people as a fraction and using a colon.	Five of the twenty-four roses in a vase are red. Write the ratio of red roses to total roses as a fraction and using a colon.
SOLUTION: The first quantity in the ratio represents the number of left-handed people, and the second quantity represents the number of all people. Ratio as a fraction $$\frac{1}{9}$$ Ratio using a colon 1:9	

Exercise 2	Exercise 3
Five of the twenty-four roses in a vase are red. Write the ratio of red roses to non-red roses as a fraction and using a colon.	Twelve out of 19 math instructors at a community college are female. Write the ratio of female math instructors at the community college to math instructors at the community college as a fraction and using a colon.
Exercise 4	**Exercise 5**
Twelve out of 19 math instructors at a community college are female. Write the ratio of male math instructors at the community college to female math instructors at the community college as a fraction and using a colon.	Nine of the 14 applicants for a job at a restaurant have previous restaurant experience. Write the ratio of applicants with experience to applicants without experience as a fraction and using a colon.

1.4.2 Simplify ratios.

A **ratio** is a comparison of two quantities with the same unit, and can be represented in multiple ways.

A ratio is simplified if the two quantities being compared share no common factors other than 1.

Simplifying a ratio is similar to simplifying a fraction to lowest terms. Divide out any factors that are common to both quantities in the ratio.

If the ratio of left-handed students to right-handed students is 4 to 36, both quantities have a common factor of 4.
Divide out the common factor, and the simplified ratio is 1 to 9.

This simplified ratio also can be expressed as 1:9 or as the fraction $\dfrac{1}{9}$.

Example	**Exercise 1**
Twenty-eight of the forty students in a math class are female. Write the ratio of female students to total students as a fraction in simplest terms.	Six of the twenty-four roses in a vase are red. Write the ratio of red roses to total roses as a fraction in simplest terms.
SOLUTION: The number of female students is 28, and the total number of students is 40. Begin with the fraction $\dfrac{28}{40}$ and simplify it to lowest terms. $$\frac{28}{40} = \frac{7}{10}$$ The ratio of female students to total students is $\dfrac{7}{10}$.	

Exercise 2
Six of the twenty-four roses in a vase are red. Write the ratio of red roses to non-red roses as a fraction in simplest terms.

Exercise 3
Six out of 54 diners at a restaurant are left-handed. Write the ratio of diners to left-handed diners as a fraction in simplest terms.

Exercise 4
Of the 3000 incoming freshmen at a college, 600 are the first person in their family to attend college. Write the ratio of first-generation students to students who are not first-generation students as a fraction in simplest terms.

Exercise 5
In a sample of 300 likely voters, 240 said they supported a bond measure to raise money for a new science building at the local college. Write the ratio of likely voters who support the bond measure to likely voters who do not support the bond measure as a fraction in simplest terms.

1.4.3   Determine unit rates.

A **rate** is a comparison of two quantities with different units, like miles and gallons.

A **unit rate** is a rate whose denominator is 1.

To convert a rate into a unit rate, divide the numerator by the denominator.

For example, suppose a package of 6 golf balls is selling for $15.

A rate comparing the cost to the number of balls is $\dfrac{\$15}{6 \text{ balls}}$.

To convert this rate to a unit rate, divide $15 \div 6$. The quotient is 2.5, so the unit rate is $2.50/ball.

The units for a unit rate are often written after the quotient, such as 55 miles/hour or $300/class.

Example	Exercise 1
If a dozen eggs cost $3.00, what is the unit cost per egg? SOLUTION: To find the unit cost, divide the total cost ($3.00) by the number of eggs (12). $$\$3.00 \div 12 = \$0.25$$ The unit cost is $0.25/egg.	A six-pack of soda cans sells for $3.09. What is the unit cost per can?

Exercise 2

A car traveled 318 miles in 5 hours. Find the average speed in miles per hour.

Exercise 3

A car traveled 400 miles using 12.8 gallons of gasoline. Find the fuel efficiency in miles per gallon.

Exercise 4

A college student can type 514 words in 8 minutes. Find the student's typing rate in words per minute.

Exercise 5

A bicycle wheel makes 450 revolutions per minute. Find the number of revolutions per second for the bicycle wheel.

1.4.4 Determine the better buy.

The **unit price** for an offer is found by dividing the price by the number of units.

For example, if a store is selling 6 golf balls for $15, the unit price is found by dividing $15 by 6. The unit price is $2.50/ball.

To determine the better buy, compute the unit price for each scenario. The offer with the lower unit price is the better buy.

Suppose that a package of 10 pens is selling for $3.00.
Its unit price is found by dividing $3.00 by 10 pens, which is $0.30/pen.

Another package of 24 pens is selling for $6.00.
Its unit price is found by dividing $6.00 by 24 pens, which is $0.25/pen.

The deal with the lower unit price is 24 pens for $6.00, so that is the better buy.

Example	Exercise 1
Determine the better buy: a 9-month gym membership for $108 or a 12-month gym membership for $129.	Determine the better buy: 12 roses for $30 or 30 roses for $60.
SOLUTION: Determine the unit cost for each membership. The better deal is the membership with the lower unit cost. 9-month membership: $$\$108 \div 9 = \$12 \,/\text{month}$$ 12-month membership: $$\$129 \div 12 = \$10.75 \,/\text{month}$$ Since the unit cost is lower for the 12-month membership, the better deal is the 12-month membership.	

Exercise 2
Determine the better buy: 2 tacos for $3.00 or 6 tacos for $7.50.

Exercise 3
Determine the better buy: 6 bottles of water for $0.79 or 24 bottles of water for $2.99.

Exercise 4
Determine the better buy: a 12-month apartment lease for $18,000 or a 6-month apartment lease for $8850.

Exercise 5
Determine the better buy: a 2-pound bag of potatoes for $2.99 or a 5-pound bag of potatoes for $7.99.

1.5.1 Solve proportions.

A **proportion** is a statement of equality between two ratios or rates.
In other words, a proportion is an equation in which two proportions are equal.

Examples of proportions:

$$\frac{5}{8} = \frac{x}{56} \qquad\qquad \frac{8}{11} = \frac{5}{x} \qquad\qquad \frac{9}{2} = \frac{4x+5}{x}$$

The **cross products** of a proportion are the two products obtained when we multiply the numerator of one fraction by the denominator of the other fraction.

The two cross products for the proportion $\frac{a}{b} = \frac{c}{d}$ are $a \cdot d$ and $b \cdot c$.

When solving a proportion, we are looking for the value of the variable that produces a true statement.
We use the fact that the cross products of a proportion are equal if the proportion is indeed true.

If $\frac{a}{b} = \frac{c}{d}$, then $a \cdot d = b \cdot c$.

To solve the proportion $\frac{5}{8} = \frac{x}{56}$, set the two cross products equal to each other and solve.

$$5 \cdot 56 = 8x$$
$$280 = 8x$$
$$35 = x$$

Example	Exercise 1
Solve $\dfrac{7}{10} = \dfrac{x}{60}$.	Solve $\dfrac{16}{30} = \dfrac{72}{x}$.

Example

Solve $\dfrac{7}{10} = \dfrac{x}{60}$.

SOLUTION:
Begin by cross multiplying, then solve the resulting equation.

$$\frac{7}{10} = \frac{x}{60}$$
$$7 \cdot 60 = 10x$$
$$420 = 10x$$
$$\frac{420}{10} = \frac{10x}{10}$$
$$42 = x$$

The solution set is $\{42\}$.

Exercise 1

Solve $\dfrac{16}{30} = \dfrac{72}{x}$.

Exercise 2

Solve $\dfrac{4}{5} = \dfrac{x}{75}$.

Exercise 3

Solve $\dfrac{39.6}{x} = \dfrac{9}{8}$.

Exercise 4

Solve $\dfrac{5}{12} = \dfrac{3x+1}{24}$.

Exercise 5

Solve $\dfrac{x+10}{40} = \dfrac{1-x}{48}$.

1.5.2 Solve applications involving proportions.

Many applications can be solved using proportions.

To set up the proportion for an applied problem, begin by identifying a ratio relating two quantities.

Select a variable to represent the unknown quantity in the problem, and set up a second ratio on the other side of the equation involving that variable.

Finish by solving the proportion.

Suppose that 2 out of every 25 students at a community college are taking a math class. If the community college has 12,000 students, how many of them are taking a math class?

The ratio of students who are taking a math class to total students at the community college is $\dfrac{2}{25}$.

The unknown is the number of students who are taking a math class. Let x represent that unknown quantity. The ratio on the other side of the proportion is $\dfrac{x}{12{,}000}$, again relating the number of students taking a math class to the total number of students at the community college.

$$\frac{2}{25} = \frac{x}{12{,}000}$$

Solve the proportion to find the number of students taking a math class.

Example	Exercise 1
One out of every nine instructors is left-handed. If a college has 225 instructors, how many would be left-handed according to this ratio?	For biology majors at a university, the ratio of female students to male students is 4:3. If the university has 52 female biology majors, how many male biology majors are there?
SOLUTION: The ratio of left-handed people to people is $\dfrac{1}{9}$. Let x represent the number of left-handed instructors out of the 225 instructors. Set up and solve the proportion. $$\frac{1}{9} = \frac{x}{225}$$ $$1 \cdot 225 = 9x$$ $$225 = 9x$$ $$\frac{225}{9} = \frac{9x}{9}$$ $$25 = x$$ According to this ratio, 25 instructors would be left-handed.	

Exercise 2

For biology majors at a university, the ratio of female students to male students is 4:3. If the university has 112 biology majors, how many are female?

Exercise 3

Approximately 2 out of 25 men are color blind. In a room of 325 men, how many would be expected to be color blind?

Exercise 4

For medium-sized parking lots, federal guidelines recommend that 1 out of every 50 parking spaces be accessible for people with disabilities. If a parking lot has 550 parking spaces, how many should be accessible spaces?

Exercise 5

The directions for a fertilizer state that 5 tablespoons of the fertilizer should be mixed with 2 gallons of water. How many tablespoons of the fertilizer should be mixed with 5 gallons of water?

1.6.1 Identify U.S. units of length, weight, and capacity.

The two common systems of measurement that are used are the U.S. system and the metric system. The U.S. system is used for measurements of length, weight, and capacity.

Common units of length in the U.S. system include inches, feet, yards, and miles.
- There are 12 inches in 1 foot.
- There are 3 feet in 1 yard.
- There are 5280 feet in 1 mile.

Common units of weight in the U.S. system include ounces, pounds, and tons.
- There are 16 ounces in 1 pound.
- There are 2000 pounds in 1 ton.

Common units of capacity in the U.S. system include fluid ounces, cups, pints, quarts, and gallons.
- There are 8 fluid ounces in 1 cup.
- There are 2 cups in 1 pint.
- There are 2 pints in 1 quart.
- There are 4 quarts in 1 gallon.

Example When measuring length, how many feet are equivalent to one mile? SOLUTION: There are 5280 feet in 1 mile.	**Exercise 1** When measuring weight, how many ounces are equivalent to one pound?

Exercise 2
When measuring capacity, how many fluid ounces are equivalent to one cup?

Exercise 3
When measuring length, how many inches are equivalent to one foot?

Exercise 4
When measuring weight, how many pounds are equivalent to one ton?

Exercise 5
When measuring capacity, how many quarts are equivalent to one gallon?

1.6.2 Perform unit conversions among U.S. units (including mixed units).

A **unit factor** is a fraction that is equal to 1 and contains different units in its numerator and denominator.

Because 12 inches = 1 foot, we can write two unit factors based on this fact:

$$\frac{12 \text{ inches}}{1 \text{ foot}} \text{ and } \frac{1 \text{ foot}}{12 \text{ inches}}$$

To convert a number of feet to inches, you can multiply the number of feet by $\frac{12 \text{ inches}}{1 \text{ foot}}$.

You can then divide the common units of feet, leaving inches as the only unit.

To convert a number of inches to feet, you can multiply the number of inches by $\frac{1 \text{ foot}}{12 \text{ inches}}$.

You can then divide the common units of inches, leaving feet as the only unit.

Suppose you wanted to convert 5 miles to feet. Begin by finding a conversion factor relating feet and miles, 5280 feet = 1 mile. Use that fact to create a unit factor with the desired units (feet) in the numerator and the initial unit (miles) in the denominator. That unit factor is $\frac{5280 \text{ feet}}{1 \text{ mile}}$. Multiply 5 miles by that unit factor to convert from miles to feet.

$$5 \text{ miles} \cdot \frac{5280 \text{ feet}}{1 \text{ mile}} = 26,400 \text{ feet}$$

Example	Exercise 1
Convert 5 feet to inches. SOLUTION: Since 12 inches are equivalent to 1 foot, multiply the quantity by the fraction $\frac{12 \text{ inches}}{1 \text{ foot}}$. $$5 \text{ feet} \cdot \frac{12 \text{ inches}}{1 \text{ foot}} = 60 \text{ inches}$$	Convert 17 pints to quarts.

Exercise 2 Convert $7\frac{1}{4}$ pounds to ounces.	**Exercise 3** Convert 10 gallons to fluid ounces.
Exercise 4 Convert 90 miles/hour to feet/second.	**Exercise 5** Convert 120 gallons/minute to fluid ounces/second.

1.6.3   Identify metric units of length, mass, and capacity.

The two common systems of measurement that are used are the U.S. system and the metric system. The metric system is used for measurements of length, mass, and capacity.

Common units of length in the metric system include millimeters, centimeters, meters, and kilometers.
- There are 1000 millimeters in 1 meter.
- There are 100 centimeters in 1 meter.
- There are 1000 meters in 1 kilometer.

Common units of mass in the metric system include milligrams, grams, and kilograms.
- There are 1000 milligrams in 1 gram.
- There are 1000 grams in 1 kilogram.

Common units of capacity in the metric system include milliliters and liters.
- There are 1000 millimeters in 1 liter.
- Millimeters are also referred to as cubic centimeters.

Example	Exercise 1
When measuring length, how many centimeters are equivalent to one kilometer? SOLUTION: Every kilometer contains 1000 meters. Each of those meters contains 100 centimeters. So, 1 kilometer is equal to $1000 \cdot 100$ centimeters, or 100,000 centimeters.	When measuring mass, how many milligrams are equivalent to one gram?

Exercise 2
When measuring capacity, how many milliliters are equivalent to one liter?

Exercise 3
When measuring length, how many meters are equivalent to one kilometer?

Exercise 4
When measuring mass, how many grams are equivalent to one kilogram?

Exercise 5
When measuring length, how many meters are equivalent to one millimeter?

1.6.4 Perform unit conversions among metric units.

A **unit factor** is a fraction that is equal to 1 and contains different units in its numerator and denominator.

Because 1000 meters = 1 kilometer, we can write two unit factors based on this fact:

$$\frac{1000 \text{ meters}}{1 \text{ kilometer}} \text{ and } \frac{1 \text{ kilometer}}{1000 \text{ meters}}$$

To convert a number of kilometers to meters, you can multiply the number of kilometers by $\frac{1000 \text{ meters}}{1 \text{ kilometer}}$. You can then divide the common units of kilometers, leaving meters as the only unit.

To convert a number of meters to kilometers, you can multiply the number of meters by $\frac{1 \text{ kilometer}}{1000 \text{ meters}}$. You can then divide the common units of meters, leaving kilometers as the only unit.

Suppose you wanted to convert 2500 milligrams to grams.

Begin by finding a conversion factor relating milligrams and grams, 1000 milligrams = 1 gram.

Use that fact to create a unit factor with the desired units (grams) in the numerator and the initial unit (milligrams) in the denominator. Multiply 2500 milligrams by that unit factor to convert from milligrams to grams.

$$2500 \text{ milligrams} \cdot \frac{1 \text{ gram}}{1000 \text{ milligrams}} = 2.5 \text{ grams}$$

Example	**Exercise 1**
Convert 8 meters to centimeters.	Convert 750 milliliters to liters.
SOLUTION: Since 1 meter is equivalent to 100 centimeters, multiply the quantity by the fraction $\frac{100 \text{ centimeters}}{1 \text{ meter}}$ to convert from meters to centimeters. $8 \text{ meters} \cdot \frac{100 \text{ centimeters}}{1 \text{ meter}} = 800 \text{ centimeters}$	

Exercise 2
Convert 1.2 kilograms to grams.

Exercise 3
Convert 1.2 grams to kilograms.

Exercise 4
Convert 108 kilometers/hour to meters/second.

Exercise 5
Convert 630 liters/hour to milliliters/second.

1.6.5 Convert between U.S. and metric units.

Conversion Factors Between the U.S. and Metric Systems

Length

English to Metric	Metric to English
1 mile ≈ 1.61 kilometers	1 kilometer ≈ 0.62 miles
1 foot ≈ 0.305 meters	1 meter ≈ 3.28 feet
1 inch = 2.54 centimeters	1 centimeter ≈ 0.39 inches

Capacity

English to Metric	Metric to English
1 gallon ≈ 3.785 liters	1 liter ≈ 0.264 gallons

Weight/Mass

English to Metric	Metric to English
1 pound ≈ 0.454 kilograms	1 kilogram ≈ 2.2 pounds
1 ounce ≈ 28.35 grams	1 gram ≈ 0.035 ounces

To convert 5 miles to kilometers, multiply by the unit factor $\dfrac{1.61 \text{ kilometers}}{1 \text{ mile}}$.

To convert 5 kilometers to miles, multiply by the unit factor $\dfrac{0.62 \text{ miles}}{1 \text{ kilometer}}$.

Example	**Exercise 1**
Convert 80 centimeters to inches. Round to the nearest tenth of an inch.	Convert 9 gallons to liters. Round to the nearest tenth of a liter.
SOLUTION: Use the conversion factor of 0.39 inches ≈ 1 centimeter .	
$80 \text{ centimeters} \cdot \dfrac{0.39 \text{ inches}}{1 \text{ centimeters}} = 31.2 \text{ inches}$	
80 centimeters is approximately equal to 31.2 inches.	

Exercise 2 Convert 125 kilograms to pounds. Round to the nearest tenth of a pound.	**Exercise 3** Convert 200 miles to kilometers. Round to the nearest tenth of a kilometer.
Exercise 4 Convert 8.3 liters to fluid ounces. Round to the nearest tenth of a fluid ounce.	**Exercise 5** Convert 142 ounces to kilograms. Round to the nearest tenth of a kilogram.

1.6.6   Solve applications involving units of measurement.

Many applications involving measurement can be solved using unit factors.

If you cannot find a single unit factor to convert from the initial unit to the desired unit, you may have to multiply by a series of unit factors.

A pickup truck has a maximum payload of 1.25 tons. How many 50-pound bags of fertilizer can it carry?
One way to solve this problem is to convert 1.25 tons into a number of 50-pound bags.

To convert 1.25 tons to pounds, multiply by the unit factor $\dfrac{2000 \text{ pounds}}{1 \text{ ton}}$.

$$1.25 \text{ tons} \cdot \frac{2000 \text{ pounds}}{1 \text{ ton}} = 2500 \text{ pounds}$$

To convert 2500 pounds to a number of bags, multiply by the unit factor $\dfrac{1 \text{ bag}}{50 \text{ pounds}}$.

$$2500 \text{ pounds} \cdot \frac{1 \text{ bag}}{50 \text{ pounds}} = 50 \text{ bags}$$

Example	Exercise 1
Each dose of the flu shot contains 0.5 milliliters of the vaccine. How many liters of the vaccine are needed in order to vaccinate 260 patients? SOLUTION: First determine how many milliliters are needed for 260 patients. $$260 \text{ patients} \cdot \frac{0.5 \text{ milliliter}}{1 \text{ patient}} = 130 \text{ milliliters}$$ Convert 130 milliliters to liters. $$130 \text{ milliliters} \cdot \frac{1 \text{ liter}}{1000 \text{ milliliters}} = 0.13 \text{ liters}$$ So, 0.13 liters of the vaccine are needed to vaccinate 260 patients.	A race organizer wants to run a rope barrier on each side of the street where a 3-mile race is being held. How many feet of rope will be needed to put the barrier on each side of the course?

Exercise 2

A box truck has a payload capacity of 2 tons. Does the truck have sufficient capacity to deliver a load of one hundred fifty 25-pound bags of rice?

Exercise 3

A carpenter needs pieces of wood that are 7 inches long. If she has a board that is 8 feet long, how many inches will be leftover after she cuts as many 7-inch pieces from the board?

Exercise 4

A gardener applies 40 ounces of fertilizer per lawn. How many lawns can be fertilized with a 50-pound bag of fertilizer?

Exercise 5

The flow rate of a river after a storm was measured to be 6000 cubic feet per second. If one cubic foot is approximately equal to 7.48 gallons, how many gallons of water would pass through in 30 minutes?

1.7.1 Classify sets of numbers.

The set of real numbers contains many subsets of numbers.

One subset is the set of **natural numbers**:
$$1, 2, 3, 4, 5, 6, 7, 8, 9, \dots$$

Another subset is the set of **whole numbers**:
$$0, 1, 2, 3, 4, 5, 6, 7, 8, 9, \dots$$
The set of whole numbers is the set of natural numbers along with the number 0.

The set of **integers** is made up of the set of whole numbers and their opposites:
$$\dots, -4, -3, -2, -1, 0, 1, 2, 3, 4, \dots$$

The set of **rational numbers** consists of numbers that can be expressed as a fraction whose numerator and denominator are integers (the denominator cannot equal 0).
Decimal numbers that terminate or repeat are rational numbers.

Irrational numbers cannot be expressed as a fraction whose numerator and denominator are integers (the denominator cannot equal 0).
Decimal numbers that do not terminate or repeat are irrational numbers.

Example	Exercise 1
To which sets of numbers does -12 belong to? (List all that apply.)	To which sets of numbers does 6 belong to? (List all that apply.)
a. Natural numbers	a. Natural numbers
b. Whole numbers	b. Whole numbers
c. Integers	c. Integers
d. Rational numbers	d. Rational numbers
e. Irrational numbers	e. Irrational numbers
SOLUTION:	
Since -12 is negative, it cannot be a natural number or a whole number.	
It is an integer, so it is also a rational number.	
Since -12 is a rational number, it is not an irrational number.	
-12 is …	
c. integer	
d. rational number	

Exercise 2
To which sets of numbers does 3.14 belong to?
(List all that apply.)
a. Natural numbers
b. Whole numbers
c. Integers
d. Rational numbers
e. Irrational numbers

Exercise 3
To which sets of numbers does π belong to?
(List all that apply.)
a. Natural numbers
b. Whole numbers
c. Integers
d. Rational numbers
e. Irrational numbers

Exercise 4

To which sets of numbers does $\frac{3}{7}$ belong to?

(List all that apply.)
a. Natural numbers
b. Whole numbers
c. Integers
d. Rational numbers
e. Irrational numbers

Exercise 5
To which sets of numbers does 0 belong to?
(List all that apply.)
a. Natural numbers
b. Whole numbers
c. Integers
d. Rational numbers
e. Irrational numbers

1.7.2 Find square roots.

The **principal square root** of b, denoted $\sqrt{b}$, is the positive number a such that $a^2 = b$.

The expression $\sqrt{b}$ is called a **radical expression**.

The sign $\sqrt{}$ is called a **radical sign**, whereas the expression contained inside is called a **radicand**.

$\sqrt{36} = 6$ because $6^2 = 36$.

$\sqrt{\dfrac{4}{9}} = \dfrac{2}{3}$ because $\left(\dfrac{2}{3}\right)^2 = \dfrac{4}{9}$.

The principal square root of a negative number, such as $\sqrt{-49}$, is not a real number because there is no real number a such that $a^2 = -49$.

Example	Exercise 1
Simplify $\sqrt{1089}$.	Simplify $\sqrt{25}$.
SOLUTION: Find the number whose square is 1089. Since $33^2 = 1089$, $\sqrt{1089} = 33$.	

Exercise 2

Simplify $\sqrt{169}$.

Exercise 3

Simplify $\sqrt{784}$.

Exercise 4

Simplify $\sqrt{\dfrac{81}{256}}$.

Exercise 5

Simplify $\sqrt{\dfrac{49}{121}}$.

1.7.3 Approximate square roots.

A calculator (or other technology) can be used to find square roots, but it can also be used to approximate square roots.

Consider the expression $\sqrt{12}$. There is no positive integer that is the square root of 12.

We know that $\sqrt{12}$ is between 3 and 4 because $\sqrt{9} = 3$, $\sqrt{16} = 4$, and 12 is between 9 and 16.

In such a case we can use a calculator to approximate the square root.

Here is the output from the Texas Instruments TI-84 calculator.

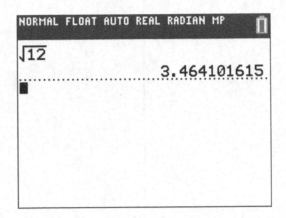

Example	Exercise 1
Approximate $\sqrt{72}$ to the nearest hundredth.	Approximate $\sqrt{2}$ to the nearest hundredth.
SOLUTION: Using a calculator, $\sqrt{72} \approx 8.49$.	

Exercise 2

Approximate $\sqrt{300}$ to the nearest hundredth.

Exercise 3

Approximate $\sqrt{0.5}$ to the nearest hundredth.

Exercise 4

Approximate $\sqrt{2654}$ to the nearest hundredth.

Exercise 5

Approximate $\sqrt{160}$ to the nearest hundredth.

1.7.4   Use the properties of real numbers.

Commutative Property

For all real numbers a and b,

$$a+b=b+a \quad \text{and} \quad a \cdot b = b \cdot a$$

The commutative property states that changing the order of the addends in a sum, or the factors in a product, does not change the result.

This property only applies to addition and multiplication, not to subtraction or division.

Associative Property

For all real numbers a, b, and c,

$$(a+b)+c = a+(b+c) \quad \text{and} \quad (a \cdot b) \cdot c = a \cdot (b \cdot c)$$

This property states that changing the grouping of addends in a sum, or factors in a product, does not change the result.

This property only applies to addition and multiplication, not to subtraction or division.

Distributive Property

For all real numbers a, b, and c,

$$a(b+c) = ab + ac$$

This property states that we can multiply the factor outside the parentheses by each term inside the parentheses.

Example	Exercise 1
Simplify $4(x-8)$ by using the distributive property.	Simplify $-5(x-19)$ by using the distributive property.
SOLUTION: Multiply the factor 4 by each term in the parentheses. $$4(x-8) = 4 \cdot x - 4 \cdot 8$$ $$= 4x - 32$$	

Exercise 2	**Exercise 3**
Rewrite $6(9x)$ using the associative property of multiplication and simplify.	Simplify $7(-6x+8)$ by using the distributive property.
Exercise 4	**Exercise 5**
Simplify $-2(8x+11)$ by using the distributive property.	Does the commutative property apply to subtraction? If not, provide an example that shows it does not apply.

1.7.5 Use the order of operations with real numbers (including grouping symbols).

The **order of operations** agreement is a standard order in which arithmetic operations are performed, ensuring a single correct result.

Order of Operations
1. **Remove grouping symbols.**
   Begin by simplifying all expressions within parentheses, brackets, and absolute value bars. Also, perform any operations in the numerator and denominator of a fraction. This is done by following Steps 2 – 4, presented next.

2. **Perform any operations involving exponents.**
   After all operations have been performed inside grouping symbols, simplify any exponential expressions.

3. **Multiply and divide.**
   These two operations have equal priority. Find products or quotients in the order that they appear from left to right.

4. **Add and subtract.**
   At this point, the only remaining operations should be addition and subtraction. Again, these operations are of equal priority, and they are performed in the order they appear from left to right.

Example	**Exercise 1**
Simplify $7 + 4 \cdot 9$.	Simplify $10(6 - 8 \cdot 11)$.
SOLUTION: Begin by finding the product, then find the sum. $$7 + 4 \cdot 9 = 7 + 36$$ $$= 43$$	

Exercise 2

Simplify $(-10)^2 - 4(3)(-5)$.

Exercise 3

Simplify $-4 + 5(7 - 3 \cdot 8)$.

Exercise 4

Simplify $3 - 6(5^2 - 4 + 3 \cdot 9)$.

Exercise 5

Simplify $18 + \left[9 - (4 - 5 \cdot 8)\right] \div 3^2$.

1.7.6 Solve applications involving real numbers.

To solve an application problem, you must decide which operation or operations are to be performed, and in what order.

Key Ideas
- Addition:
  Problems involving a total of two or more quantities

- Subtraction:
  Problems involving the difference between two quantities, or how much remains after one quantity is taken away from another quantity

- Multiplication:
  Problems involving one quantity being used multiple times

- Division:
  Problems involving breaking up a quantity into equally sized quantities

Example	Exercise 1
A group of 4 friends went out to dinner. If each person paid $40, what was the total bill? SOLUTION: To determine the total bill, multiply the number of friends by the amount that each paid. $$4(\$40) = \$160$$ The total bill was $160.	Three friends started a food-truck business. Last month the business made a profit of $1350. If the friends reinvest half of their profits in the business and split the remaining profit evenly, how much money did each of the 3 friends receive last month?

Exercise 2

Tina owns 200 shares of a stock that dropped in value by $4.35 per share last year. She also owns 500 shares of a stock that increased in value by $16.71 last year. What was Tina's net income on these two stocks last year?

Exercise 3

A mother gave birth to twins. One twin weighed $5\frac{3}{8}$ pounds and the other weighed $4\frac{11}{16}$ pounds. What was the total weight of the twins?

Exercise 4

A manufacturer recommends using 1.25 fluid ounces of weed spray concentrate per gallon of water. How many fluid ounces of weed spray concentrate should be mixed with 12 gallons of water?

Exercise 5

An office manager bought 8 cases of paper at a price of $34.99 per case (tax included). If she paid with three $100 bills, how much change did she get back?

2.1.1 Evaluate algebraic expressions.

To **evaluate** an algebraic expression with one variable for a particular value, substitute the value for the variable and simplify the resulting expression using the order of operations.

Replace the variable with a set of parentheses, then write the value inside the parentheses. At that point you can simplify the expression.

Evaluating the expression $3x + 25$ for $x = 6$ asks you to determine what $3x + 25$ is equal to when x is 6.

$$3x + 25$$
$$3(6) + 25$$
$$= 18 + 25$$
$$= 43$$

When evaluating an expression that has more than one variable, substitute the appropriate values for each variable and simplify.

Example	Exercise 1
Evaluate $-2x + 37$ for $x = 6$.	Evaluate $8x - 19$ for $x = -9$.
SOLUTION: Substitute 6 for x and simplify using the order of operations. $$-2x + 37$$ $$-2(6) + 37$$ $$= -12 + 37$$ $$= 25$$	

Exercise 2

Evaluate $x^2 - 5x - 14$ for $x = 10$.

Exercise 3

Evaluate $x^2 + 3x - 40$ for $x = -7$.

Exercise 4

Evaluate $9a - 4b$ for $a = -4$ and $b = 17$.

Exercise 5

Evaluate $b^2 - 4ac$ for $a = 2$, $b = -8$, and $c = 11$.

2.1.2 Combine like terms.

In an algebraic expression, a **term** is a number, a variable, or a product of a number and variable factors.

Two terms that have the same variable factors with the same exponents, or that are both constants, are called **like terms**.

Combining Like Terms

When simplifying variable terms, combine like terms into a single term with the same variable part by adding or subtracting the coefficients of the like terms.

Consider the expression $13a - 5b + 2c + a + 3b$.
- There are two terms whose variable factor is a.
  Combine those two terms by combining their two coefficients, 13 and 1: $13a + a = 14a$.
- There are two terms whose variable factor is b.
  Combine those two terms by combining their two coefficients, -5 and 3: $-5b + 3b = -2b$.
- There is only one term with the variable factor c, and it cannot be combined with any other term.
$$13a - 5b + 2c + a + 3b = 14a - 2b + 2c$$

Example	Exercise 1
Simplify $7x - 15x$. SOLUTION: Combine like terms by adding the coefficients: $7 + (-15) = -8$. $$7x - 15x = -8x$$	Simplify $22x + 45x$.

Exercise 2
Simplify $-9m - 32m$.

Exercise 3
Simplify $4x + 13 - x + 55$.

Exercise 4
Simplify $2x - 3y - 4x - 11y$.

Exercise 5
Simplify $5a + b - 4c - 12 - 2a + 7b - 12$.

2.1.3 Simplify algebraic expressions.

To simplify a variable expression, begin by applying the distributive property if possible.

Then combine any like terms.

If you needed to simplify the expression $8(3x-4)+5x$, begin by distributing the factor 8 to the two terms inside the parentheses. Then combine any like terms.

$$8(3x-4)+5x = 8 \cdot 3x - 8 \cdot 4 + 5x$$
$$= 24x - 32 + 5x$$
$$= 29x - 32$$

When simplifying the expression $2(9x-11)-7(4x-18)$, be sure to distribute the negative factor -7 to each term in the second set of parentheses.

$$2(9x-11)-7(4x-18) = 18x - 22 - 28x + 126$$
$$= -10x + 104$$

Example	Exercise 1
Simplify $3(2x-15)-7$.	Simplify $8x+3y-x-14y$.
SOLUTION: Begin by applying the distributive property, then combine any like terms. $$3(2x-15)-7 = 6x - 45 - 7$$ $$= 6x - 52$$	

Exercise 2

Simplify $9(8x-7)+6x$.

Exercise 3

Simplify $5(4x-9)-6x+33$.

Exercise 4

Simplify $8-19x-4(3x-8)$.

Exercise 5

Simplify $12(2x-9)-7(4x-13)$.

2.1.4   Translate English phrases into algebraic expressions.

One important skill for setting up applied problems is the ability to translate English phrases into algebraic expressions.

You should become familiar with key phrases used to indicate addition, subtraction, multiplication, and division.

Key Phrases for Addition
- Sum, plus, increased by, more than, total

Key Phrases for Subtraction
- Difference, minus, decreased by, less than

Key Phrases for Multiplication
- Product, times, multiplied by, of, twice

Key Phrases for Division
- Quotient, divided by, ratio, split into equal parts

Use caution with the phrase less than. Seven less than a number translates to $x - 7$, not $7 - x$. Seven less than a number indicates that you are to subtract 7 from the number.

Example	Exercise 1
Build a variable expression for 8 less than a number. (Use x to represent the number.)	Build a variable expression for a number increased by 15. (Use x to represent the number.)
SOLUTION: To express 8 less than a number x, subtract 8 from x: $$x - 8$$	

Exercise 2
Build a variable expression for twice a number. (Use *x* to represent the number.)

Exercise 3
Build a variable expression for a number divided by 5. (Use *x* to represent the number.)

Exercise 4
Build a variable expression for 13 less than 4 times a number. (Use *x* to represent the number.)

Exercise 5
Build a variable expression for 7 times the difference between a number and 6. (Use *x* to represent the number.)

2.2.1   Distinguish between expressions and equations.

An **expression** is a combination of constants or variables using arithmetic operations.

An expression can involve just constants like $9+7\cdot 3$ or $(-8)^2 -4(5)(-6)$, or it can also involve variables like $\frac{2}{3}x-6$ or $x^2 -5x-14$.

Other examples of expressions:

$$3x^3 -2x^2 +5x \qquad\qquad 7\cdot 8-5\cdot 6 \qquad\qquad \frac{x-6}{x^2 -9x+18}$$

An **equation** is a mathematical statement of equality between two expressions.

An equation states that the expression on the left side of the equation is equivalent to the expression on the right side of the equation.

While an equation will typically contain variables, an equation can be made with two equivalent numerical expressions like $5\cdot 8 = 6^2 +2^2$.

Examples of equations:

$$3x-7=8 \qquad\qquad x^2 +15x+56=0 \qquad\qquad 2x^2 -7x-10 = x^2 +20$$

Example	**Exercise 1**
Is the following an expression or an equation? $$2x-4$$ SOLUTION: Since there is no equal sign stating that two expressions are equal to each other, this is an expression.	Is the following an expression or an equation? $$2x=4$$

Exercise 2
Is the following an expression or an equation?
$$x^2 + 7x = 8$$

Exercise 3
Is the following an expression or an equation?
$$x^2 + 7x + 8$$

Exercise 4
Is the following an expression or an equation?
$$2(3x + 5) - 7$$

Exercise 5
Is the following an expression or an equation?
$$2(3x + 5) - 7 = 13$$

2.2.2 Solve linear equations in one variable using the addition property of equality.

The **addition property of equality** states that the same number can be added to both sides of an equation without affecting the equality of both sides. The same is true when subtracting the same number from both sides of an equation.

Addition Property of Equality
For any expressions A and B and any number n, if $A = B$ then

$$A + n = B + n \text{ and } A - n = B - n$$

The addition property of equality can be used to isolate a variable on one side of an equation when the expression on that side has a variable with a constant added to it or subtracted from it.

For example, consider the equation $x - 17 = 45$. Adding 17 to both sides of the equation will isolate the variable x on the left side of the equation. The resulting constant on the right side of the equation is the solution of the equation.

$$x - 17 = 45$$
$$x - 17 + 17 = 45 + 17 \qquad \textit{Add 17 to both sides of the equation.}$$
$$x = 62 \qquad \textit{Simplify.}$$

The solution(s) of an equation can be written as a solution set by listing them inside a pair of curly brackets. The solution set for this equation is $\{62\}$.

Example	**Exercise 1**
Solve. $$x - 9 = -13$$ SOLUTION: Since 9 is being subtracted from x on the left side of the equation, adding 9 to both sides of the equation will isolate the variable x on the left side of the equation. $$x - 9 = -13$$ $$x - 9 + 9 = -13 + 9$$ $$x = -4$$ The solution set for this equation is $\{-4\}$.	Solve. $$x + 8 = 27$$

Exercise 2
Solve.
$$x - 45 = 62$$

Exercise 3
Solve.
$$x - \frac{7}{6} = \frac{3}{2}$$

Exercise 4
Solve.
$$-9 + x = -40$$

Exercise 5
Solve.
$$-13 = x + 33$$

2.2.3   Solve linear equations in one variable using the multiplication property of equality.

The **multiplication property of equality** states that the both sides of an equation can be multiplied by the same nonzero number without affecting the equality of both sides. The same is true when dividing both sides of an equation by the same nonzero number.

Multiplication Property of Equality
For any expressions A and B and any nonzero number n, if $A = B$ then

$$n \cdot A = n \cdot B \text{ and } \frac{A}{n} = \frac{B}{n}$$

Note that the multiplication property of equality only holds when multiplying (or dividing) both sides of an equation by a number that is not equal to 0.

For example, consider the equation $4x = -36$. Dividing both sides of the equation by 4 will isolate the variable x on the left side of the equation. The resulting constant on the right side of the equation is the solution of the equation.

$$4x = -36$$
$$\frac{4x}{4} = \frac{-36}{4} \qquad \textit{Divide both sides of the equation by 4.}$$
$$x = -9 \qquad \textit{Simplify.}$$

The solution set for this equation is $\{-9\}$.

Example	**Exercise 1**
Solve. $$-3x = 48$$	Solve. $$5x = -35$$
SOLUTION: Since the variable x is being multiplied by -3 on the left side of the equation, dividing both sides of the equation by -3 will isolate the variable x on the left side of the equation. $$-3x = 48$$ $$\frac{-3x}{-3} = \frac{48}{-3}$$ $$x = -16$$ The solution set for this equation is $\{-16\}$.	

Exercise 2
Solve.

$$12x = 200$$

Exercise 3
Solve.

$$-6x = -40$$

Exercise 4
Solve.

$$\frac{x}{7} = 77$$

Exercise 5
Solve.

$$\frac{3}{4}x = -75$$

2.2.4 **Solve linear equations in one variable using both properties of equality.**

Some equations will require you to use both the addition and multiplication properties of equality to solve them. Suppose you had to solve the equation $5x - 15 = 20$. Should you begin by applying the addition property of equality (add 15 to both sides of the equation) or by applying the multiplication property of equality (divide both sides of the equation by 5)? Begin by adding 15 to both sides of the equation to isolate the variable term, then you can divide both sides of the resulting equation by 5 to isolate the variable.

$$5x - 15 = 20$$
$$5x = 35 \qquad \text{\textit{Add 15 to both sides of the equation.}}$$
$$x = 7 \qquad \text{\textit{Divide both sides of the equation by 5 and simplify.}}$$

The solution set for this equation is $\{7\}$.

Solving Linear Equations
1. **Simplify each side of the equation completely.**
   Apply the distributive property to clear parentheses. Clear fractions by multiplying both sides of the equation by the LCM of the denominators. Combine like terms on the same side of the equation.
2. **Collect all variable terms on one side of the equation.**
   Use the addition property of equality to collect all variables on one side of the equation.
3. **Collect all constant terms on the other side of the equation.**
   Use the addition property of equality to isolate the variable term on one side of the equation.
4. **Divide both sides of the equation by the coefficient of the variable term.**
   Use the multiplication property of equality to isolate the variable and solve the equation.
5. **Check your solution.**

Example	**Exercise 1**
Solve.	Solve.
$$6x - 11 = -29$$	$$-4x + 13 = -9$$
SOLUTION:	
Since there is only one variable term, we can isolate the variable term by adding 11 to both sides of the equation.	
The equation can then be solved by dividing both sides of the equation by 6.	
$$6x - 11 = -29$$	
$$6x - 11 + 11 = -29 + 11$$	
$$6x = -18$$	
$$\frac{6x}{6} = \frac{-18}{6}$$	
$$x = -3$$	
The solution set for this equation is $\{-3\}$.	

Exercise 2
Solve.
$$3x - 8 = 5x + 20$$

Exercise 3
Solve.
$$4(2x + 11) - 35 = 81$$

Exercise 4
Solve.
$$6x - (2x + 3) = -15$$

Exercise 5
Solve.
$$\frac{5}{6}x + 3 = -\frac{2}{3}x - 15$$

2.2.5 Translate sentences into equations.

Often you will be required to translate an English sentence into an equation in order to be able to solve an applied problem. One key is to be able to identify phrases that indicate the four arithmetic operations of addition, subtraction, multiplication, and division.

Key Phrases
    Addition: sum, plus, increased by, more than, total
    Subtraction: difference, minus, less than, decreased by
    Multiplication: product, times, twice
    Division: quotient, divided by, ratio

Be careful with the phrase "less than". Some students translate "8 less than a number" to be $x - 8$ while others translate it to be $8 - x$. These two expressions are not the same. In fact, they are opposites of each other. Keep in mind that 8 less than a number is found by subtracting 8 from the number, so the correct translation is $x - 8$.

Recall that an equation is a statement of equality between two expressions. So, in addition to being able to identify the expressions, you must be able to determine the location to place the equal sign. Often the word "is" will be included in the sentence you are translating, and if it can be taken to mean "is equal to," "is equivalent to," or "is the same as," then that will be the place to insert the equal sign.

Example	**Exercise 1**
Translate the sentence to an equation and solve.	Translate the sentence to an equation and solve.
Fifteen more than twice a number is 39. Find the number.	Nine more than a number is -37. Find the number.
SOLUTION: Fifteen more than an expression means that we need to add 15 to the expression.	
Twice a number means 2 times a number. If the variable used to represent the number is x, then twice a number can be expressed as $2x$.	
Fifteen more than twice a number: $2x + 15$	
The equation is $2x + 15 = 39$.	

Exercise 2
Translate the sentence to an equation and solve.

Eight times a number is 512. Find the number.

Exercise 3
Translate the sentence to an equation and solve.

Thirty less than 4 times a number is 78. Find the number.

Exercise 4
Translate the sentence to an equation and solve.

Nine more than $\frac{3}{4}$ times a number is 27. Find the number.

Exercise 5
Translate the sentence to an equation and solve.

When 3.2 is added to 6 times a number, the result is 56.6. Find the number.

2.2.6 Solve applications involving linear equations in one variable.
Perimeter of a Rectangle

One common application of linear equations in one variable involves the perimeter of a rectangle. The formula for the perimeter of a rectangle is $2(\text{Length}) + 2(\text{Width}) = \text{Perimeter}$.

Create variable expressions for the length and the width in terms of the same variable. Try letting x represent the quantity you know the least about, and build up the other quantity in terms of x. Substitute these expressions for the length and the width, as well as the perimeter, in the formula and solve for x. Then return to your expressions for the length and width and compute those quantities for the value of x.

Sums of Consecutive Integers

Consecutive integers are integers that are next to each other on a number line, like 3, 4, and 5. If you let x represent the first integer, then the second can be represented by $x + 1$ because it is 1 larger than the first integer. The third consecutive integer can be represented by $x + 2$, the fourth by $x + 3$, and so on. The key idea is that consecutive integers increase in steps of 1.

If a problem involves the sum of consecutive *odd* integers, like 7, 9, and 11, note that these increase in steps of 2. If x represents the first odd integer, the others can be represented by $x + 2$, $x + 4$, $x + 6$, and so on. The same is true for consecutive *even* integers because they increase in steps of 2 as well.

Add the variable expressions and set that sum equal to the sum given in the problem and solve for x. Finish by identifying each integer, not just the first integer.

Example	Exercise 1
The length of a rectangle is 5 feet more than 3 times its width. The perimeter of the rectangle is 42 feet. Find the length and width of the rectangle.	The width of a rectangle is 12 inches less than its length. The perimeter of the rectangle is 56 inches. Find the length and width of the rectangle.
SOLUTION: Less is known about the width, so let x represent the width. The length is 5 feet more than 3 times the width, so it can be represented as $3x + 5$. The formula for the perimeter of a rectangle is $2(\text{Length}) + 2(\text{Width}) = \text{Perimeter}$. Substitute $3x + 5$ for the length, x for the width, and 42 for the perimeter. Solve for x. $$2(3x + 5) + 2(x) = 42$$ $$6x + 10 + 2x = 42$$ $$8x + 10 = 42$$ $$8x = 32$$ $$x = 4$$ The width (x) is 4 feet, and the length is $3(4) + 5$ or 17 feet.	

Exercise 2
One side of a triangle is 4 inches longer than the shortest side. The third side of the triangle is twice as long as the shortest side. The perimeter is 52 inches. Find the length of each side of the triangle.

Exercise 3
The sum of four consecutive integers is 418. Find them.

Exercise 4
The sum of three consecutive even integers is 384. Find them.

Exercise 5
The sum of three consecutive odd integers is 177. Find them.

2.3.1 Write inequality statements using real numbers and inequality symbols.

For any real number a, the inequality $x > a$ is used to represent all values of x that are greater than a. The inequality $x < a$ is used to represent all values of x that are less than a.

Inequalities of the form $x > a$ or $x < a$ are strict inequalities because they involve only values that are strictly greater than a $(x > a)$ or strictly less than a $(x < a)$.

The inequality $x \geq a$ represents all values of x that are greater than or equal to a.
The inequality $x \leq a$ represents all values of x that are less than or equal to a.
Inequalities of the form $x \geq a$ and $x \leq a$ are called weak inequalities because the endpoint a is included as a solution.

To translate a statement into an inequality using real numbers and inequality symbols you must be familiar with different phrases that are associated with each type of inequality.

$x > a$: greater than a, more than a, higher than a, above a
$x \geq a$: at least a, a or higher
$x < a$: less than a, lower than a, below a
$x \leq a$: at most a, a or lower

Example	Exercise 1
Write the statement as an inequality: x is less than 4. SOLUTION: The phrase x is less than 4 can be expressed as $x < 4$.	Write the statement as an inequality: x is greater than -9.

Exercise 2
Write the statement as an inequality: *x* is no more than −5.

Exercise 3
Write the statement as an inequality: *x* is 7 or greater.

Exercise 4
Write the statement as an inequality: *x* is at least −6.

Exercise 5
Write the statement as an inequality: *x* is negative.

2.3.2   Graph linear inequalities in one variable on a number line.

When graphing a linear inequality in one variable on a number line, start by writing it so that the variable is on the left side of the inequality.

To graph an inequality on a number line, begin by placing a point at the endpoint.
- If the inequality is a strict inequality involving the symbol $<$ or $>$, use an open circle at the endpoint.

- If the inequality is a weak inequality involving the symbol $\leq$ or $\geq$, use a closed circle at the endpoint.

Then shade the portion of the number line that contains the solutions.
- Solutions of $x < a$ or $x \leq a$ can be found to the left of a on the number line.

- Solutions of $x > a$ or $x \geq a$ can be found to the right of a on the number line.

Be sure to draw an arrow to indicate that the solutions continue in that direction.

Example	**Exercise 1**
Present the solutions on a number line: $$x \leq 4$$	Present the solutions on a number line: $$x > -2$$
SOLUTION: Since the endpoint 4 is included as a solution, place a closed circle at 4 on the number line. The values of x that are less than 4 can be found to the left of 4 on the number line, so shade to the left of 4.	

Exercise 2
Present the solutions on a number line:
$$x < 5$$

Exercise 3
Present the solutions on a number line:
$$x \geq 7$$

Exercise 4
Present the solutions on a number line:
$$x \leq -6$$

Exercise 5
Present the solutions on a number line:
$$x > 0$$

2.3.3 Write solutions to inequalities in set-builder notation.

In addition to graphing the solutions of an inequality, we can present the solutions of an inequality using **set-builder** notation.

Begin by writing $\{x| \quad \}$. This is read as "the set of all real numbers x, such that ..."
- The curly brackets are used to denote a set.
- The vertical bar, |, is read as "such that" and separates the variable x from the condition that defines it.

In the space that follows the vertical bar, write the inequality.
Be sure to determine whether the endpoint is included $(\leq, \geq)$ or not $(<, >)$.

- The endpoint 2 is not included and the number line is shaded to the left of 2.
  This indicates that x is less than 2: $\{x|x<2\}$

- The endpoint 2 is included and the number line is shaded to the left of 2.
  This indicates that x is less than or equal to 2: $\{x|x\leq2\}$

- The endpoint 2 is not included and the number line is shaded to the right of 2.
  This indicates that x is greater than 2: $\{x|x>2\}$

- The endpoint 2 is included and the number line is shaded to the right of 2.
  This indicates that x is greater than or equal to 2: $\{x|x\geq2\}$

Example	Exercise 1
Write the given interval in set-builder notation.	Write the given interval in set-builder notation.

SOLUTION: Note that the endpoint 8 is not included as a solution, so the inequality will be a strict inequality with no equal sign.	
The number line is shaded to the left of 8, indicating values that are less than 8.	
The interval can be expressed in set-builder notation as $\{x|x<8\}$.	

Exercise 2
Write the given interval in set-builder notation.

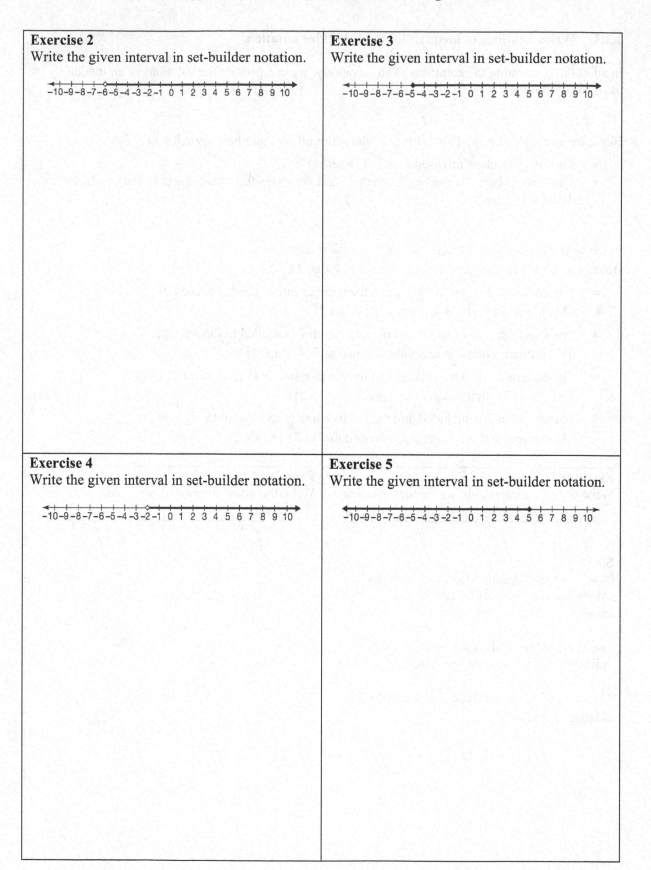

Exercise 3
Write the given interval in set-builder notation.

Exercise 4
Write the given interval in set-builder notation.

Exercise 5
Write the given interval in set-builder notation.

2.3.4   Write solutions to inequalities in interval notation.

In addition to graphing the solutions of an inequality and presenting them using set-builder notation, we can also present the solutions of an inequality using **interval** notation.

A range of values on a number line is called an interval.
We can express the interval by listing its left and right endpoints with a comma between them.

If the interval includes an arrow that indicates the solutions continue in one direction, then the interval is technically lacking an endpoint.
- If the left arrow is shaded, indicating that the solutions continue without bound in the negative direction, write the symbol $-\infty$ (negative infinity) in place of the left endpoint.
- If the right arrow is shaded, indicating that the solutions continue without bound in the positive direction, write the symbol ∞ (infinity) in place of the right endpoint.

We use parentheses or brackets to enclose the interval notation.
- If an endpoint has a closed circle on a number line, indicating the endpoint is included as a solution, then a square bracket is written next to that endpoint.
- If an endpoint has an open circle on a number line, indicating the endpoint is not included as a solution, then a parenthesis is written next to that endpoint.
- If the left or right arrow are shaded on a number line, then a parenthesis is written next to either $-\infty$ or ∞.

Example	Exercise 1
Write the given interval in interval notation.	Write the given interval in interval notation.

Example (continued)

SOLUTION:
Note that the arrow on the left side of the number line is shaded. This indicates that the left endpoint will be replaced by the symbol $-\infty$.

The right endpoint is 8, so the interval notation will be of the form $-\infty$, 8.

Since neither "endpoint" has a closed circle, the interval notation will be enclosed in parentheses.

Interval notation: $(-\infty, 8)$

Exercise 2
Write the given interval in interval notation.

Exercise 3
Write the given interval in interval notation.

Exercise 4
Write the given interval in interval notation.

Exercise 5
Write the given interval in interval notation.

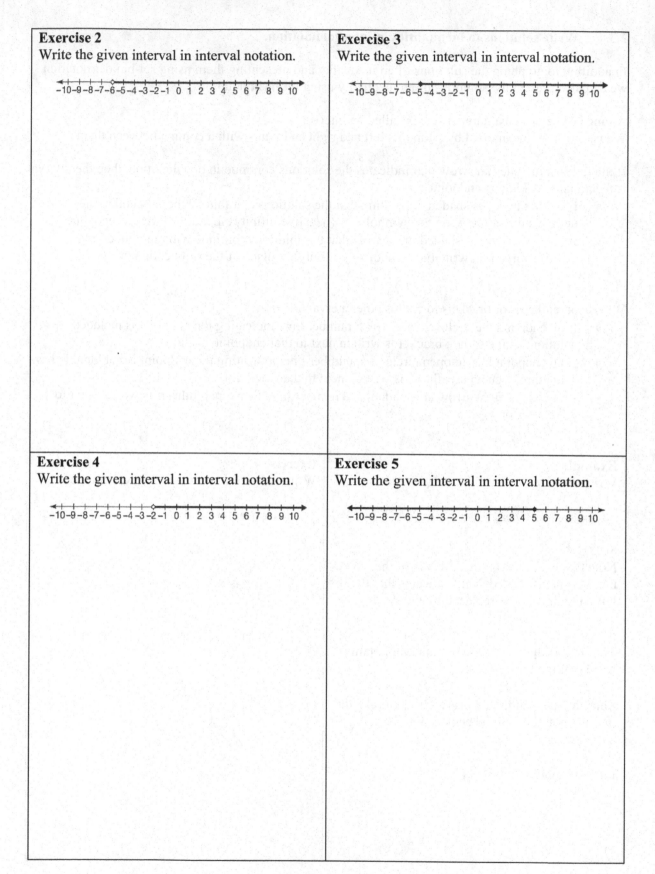

2.3.5 Solve linear inequalities in one variable.

Solving linear inequalities in one variable is similar to solving linear equations in one variable, with one exception.

> Whenever you multiply or divide both sides of an inequality by a negative number, the direction of the inequality sign changes.

It will be to your advantage if you isolate the variable on the left side of the inequality because it will help you to determine which part of the number line to shade.

> **Solving Linear Inequalities**
> 1. **Simplify each side of the inequality completely.**
> 2. **Collect all variable terms on one side of the inequality.**
>   When it comes time to graphing solutions on a number line, having the variable on the left side of the inequality can help with shading.
> 3. **Collect all constant terms on the other side of the inequality.**
> 4. **Divide both sides of the inequality by the coefficient of the variable term.**
>   If you divide both sides of the inequality by a negative number, be sure to change the direction of the inequality sign.
> 5. **Express your solutions in the desired format.**
>   This could be on a number line, using set-builder notation, or using interval notation.

Example	**Exercise 1**
Solve $-2x+9>15$.	Solve $3x-8\le 7$.
Express the solutions using interval notation.	Express the solutions using interval notation.
SOLUTION:	
Begin by isolating the variable term on the left side of the inequality. This can be done by subtracting 9 from both sides of the inequality.	
At that point the coefficient of the variable term is negative, so when dividing by -2 the direction of the inequality sign will have to be reversed.	
$$-2x+9>15$$ $$-2x>6$$ $$\frac{-2x}{-2}<\frac{6}{-2}$$ $$x<-3$$	
The graph of the solutions on the number line would have an open circle at -3 and be shaded to the left of -3.	
Interval notation: $(-\infty,\ -3)$	

Exercise 2
Solve $-4x+15>23$.
Express the solutions using interval notation.

Exercise 3
Solve $5x-6>2x+15$.
Express the solutions using interval notation.

Exercise 4
Solve $2x+9 \geq 4x+25$.
Express the solutions using interval notation.

Exercise 5
Solve $2(3x+5)-1 \leq -9$.
Express the solutions using interval notation.

2.3.6   Translate sentences into linear inequalities in one variable.

To translate sentences into inequalities, you must be familiar with the key words associated with inequalities.

Key Words for Inequalities
$x > a$: greater than a, more than a, higher than a, above a
$x \geq a$: at least a, a or higher
$x < a$: less than a, lower than a, below a
$x \leq a$: at most a, a or lower

In addition, you will often have to be able to translate an English phrase into an expression. Here is a reminder of those key phrases needed.

Key Phrases for Building Expressions
Addition: sum, plus, increased by, more than, total
Subtraction: difference, minus, less than, decreased by
Multiplication: product, times, twice
Division: quotient, divided by, ratio

Example	Exercise 1
A student must get at least 8 questions correct in order to pass the quiz. Write an inequality that shows the number of questions that must be correct in order for a student to pass the quiz. SOLUTION: Let x represent the number of questions that are correct. Since a student must get at least 8 questions to pass the quiz, the inequality associated with passing the quiz is $x \geq 8$.	An airport shuttle can fit up to 10 passengers. Write an inequality that shows the number of passengers that the shuttle can fit.

Exercise 2 A doctor tells her patient to head to the emergency room if his temperature is above 102°F. Write an inequality that shows the temperatures for which the patient is supposed to go to the emergency room.	**Exercise 3** A hotel with 20 vacancies will break even on a given night. Write an inequality that shows the number of vacancies for which the hotel will make a profit.
Exercise 4 A veterinarian's office has enough supplies to spay 8 cats. Write an inequality that shows the number of cats for which the office does not have enough supplies.	**Exercise 5** A restaurant hosts business meetings, charging $200 plus $25 for each attendee. Michelle is planning a business meeting and has a budget of $500. Write an inequality that can be used to determine the number of attendees that can be invited without going over budget.

2.3.7 Solve applications involving linear inequalities in one variable.

The key to solving an application involving a linear inequality is to be able to set up the linear expression related to the given scenario.

If a particular item has a value of m dollars, then the total value of x of those items is mx.
For example, if a company is selling smart phone apps for $2 then the revenue earned from selling x apps is $2x$.

Some quantities will have an added fixed value.
For example, suppose you are attending the county fair. If it costs $10 to get into the fair plus $4 per ride, the total cost for going on x rides is $4x + 10$. The $4x$ represents the cost of riding x rides, and the additional $10 is the admission fee.

The average of a set of n values can be found by finding the sum of the n values and dividing by n. For example, suppose Marissa has scored 96, 87, and 95 on the first three exams in a math class. Her average score after the fourth exam could be given by the expression $\dfrac{96 + 87 + 95 + x}{4}$, where x is her score on the fourth exam.

Example	Exercise 1
Kevin earns a commission of $400 for every used car he sells. How many cars does he have to sell to earn more than $12,500 in commissions this month?	A company is trying to raise capital by selling 2 million shares of stock. What price does it need to receive per share in order to raise at least $70 million?
SOLUTION: Since Kevin earns $400 for each sale he makes, the total commission for making x sales is given by the expression $400x$. To determine how many cars he must sell to earn more than $12,500, we must solve the inequality $400x > 12,500$. The first step is to divide both sides of the inequality by 400. $$400x > 12,500$$ $$\frac{400x}{400} > \frac{12,500}{400}$$ $$x > 31.25$$ Since the number of cars sold must be a whole number, Kevin will have to sell 32 cars to earn more than $12,500.	

Exercise 2

A restaurant hosts business meetings, charging $200 plus $25 for each attendee. Michelle is planning a business meeting and has a budget of $500. How many attendees can Michelle invite without going over budget?

Exercise 3

A business traveler has an allowance of $100 for a rental car. She can rent a luxury car for $49.95 plus $0.10 per mile. How many miles can the traveler drive without exceeding her allowance?

Exercise 4

In order to qualify for a state bowling tournament, Tim must have an average score of at least 200 per game in a 3-game qualifier. If Tim scored 225 and 189 in the first two games, what scores in the third game will make him eligible for the state tournament?

Exercise 5

Students with an average test score below 70 after the fourth quiz will be sent an Early Alert warning. Robert scored 72, 63, and 50 on the first three quizzes. What scores on the fourth exam will save Robert from receiving an Early Alert warning?

2.4.1 Solve a formula for a specific variable.

Solving a formula for a specified variable is similar to solving a linear equation.

For example, solving the formula $5x + 4y = 60$ for x is similar to solving the equation $5x + 4 = 60$.

Treat each term containing a variable other than the variable you are solving for as if it were a constant term.

Isolate the term containing the variable you are solving for on one side of the equation, and collect all constant terms as well as terms containing other variables on the other side of the equation. Use the addition property of equality to achieve this.

If the isolated variable term has a constant factor or other variables, use the multiplication property of equality to isolate the desired variable.

So, when solving the formula $5x + 4y = 60$ for x, first isolate the variable term $5x$ on the left side by subtracting $4y$ from both sides of the equation.
$$5x = -4y + 60$$

Then finish solving for x by dividing both sides of the equation by 5.
$$x = \frac{-4y + 60}{5}$$

Example	Exercise 1
Solve for y: $3x + 2y = 6$.	Solve for y: $-4x + 3y = 19$.
SOLUTION: Isolate the term containing the variable y by subtracting $3x$ from both sides of the equation. To isolate the variable y, finish by dividing both sides of the equation by 2. $$3x + 2y = 6$$ $$2y = -3x + 6$$ $$y = \frac{-3x + 6}{2}$$	

Exercise 2
Solve for y: $x + 8y = -10$.

Exercise 3
Solve for y: $3x + 4y + 5z = 12$.

Exercise 4
Solve for L: $P = 2L + 2W$

Exercise 5
Solve for h: $A = 2\pi rh + 2\pi r^2$

2.4.2 Find the perimeter of a figure.

The perimeter of a figure composed of straight sides is the distance around the outside of the figure, and is equal to the sum of all its sides.

If each side is measured in the same unit, the perimeter is expressed in terms of the same unit. For example, if each side is measured in inches, the perimeter is expressed in inches as well.

Certain figures have formulas for their perimeters.

Rectangle with length L and width W:
$$\text{Perimeter} = 2(\text{Length}) + 2(\text{Width})$$
or
$$P = 2L + 2W$$

Square with side x:
$$\text{Perimeter} = 4(\text{Side})$$
or
$$P = 4x$$

Triangle with sides a, b, and c:
$$P = a + b + c$$

Example	Exercise 1
Find the perimeter of a rectangle whose length is 17 inches and whose width is 11 inches.	Find the perimeter of a rectangle whose length is 27 centimeters and whose width is 12 centimeters less than its length.
SOLUTION: The formula for the perimeter of a rectangle is $\text{Perimeter} = 2(\text{Length}) + 2(\text{Width})$ or $P = 2L + 2W$.	
Substitute 17 for the length and 11 for the width.	
$$\begin{aligned} P &= 2L + 2W \\ P &= 2(17) + 2(11) \\ P &= 34 + 22 \\ P &= 56 \end{aligned}$$	
The perimeter is 56 inches.	

Exercise 2
Find the perimeter of a triangle whose three sides have length 3 feet, 4 feet, and 6 feet.

Exercise 3
Find the perimeter of an equilateral triangle that has sides with lengths of 8 millimeters.

Exercise 4
Find the perimeter of a square that has a side of length $6\frac{3}{5}$ inches.

Exercise 5
A regular octagon is an 8-sided figure whose eight sides all have equal length. Find the perimeter of a regular octagon that has a side of 13.25 inches.

2.4.3   Find the circumference of a circle.

The **circumference** of a circle is the distance around the outside of the circle, and is based on the circle's radius (distance from the center of the circle to a point on the circle) or diameter (length of a segment between two points on the circle that passes through the center of the circle).

The circumference of a circle is computed in the same units its radius (or diameter) is measured.

Circumference of a circle with radius r:
$$\text{Circumference} = 2\pi \cdot \text{Radius}$$
$$\text{or}$$
$$C = 2\pi r$$

Circumference of a circle with diameter d:
$$\text{Circumference} = \pi \cdot \text{Diameter}$$
$$\text{or}$$
$$C = \pi d$$

π is an irrational number that is approximately equal to 3.14159. When approximating the circumference of a circle, use the built-in value of π on your calculator.

Example	Exercise 1
Find the circumference of a circle whose radius is 8 inches. Round to the nearest tenth of an inch.	Find the circumference of a circle whose radius is 15 centimeters. Round to the nearest tenth of a centimeter.
SOLUTION: The formula for the circumference of a circle is $\text{Circumference} = 2\pi \cdot \text{Radius}$ or $C = 2\pi r$. Substitute 8 for the radius. $$C = 2\pi r$$ $$C = 2\pi(8)$$ $$C = 16\pi$$ The exact circumference is 16π inches. Using a calculator's built in value for π, this is approximately 50.3 inches.	

Exercise 2
Find the circumference of a circle whose diameter is 12 feet. Round to the nearest tenth of a foot.

Exercise 3
Find the circumference of a circle whose diameter is 40 millimeters. Round to the nearest tenth of a millimeter.

Exercise 4
Find the circumference of a circle whose radius is 2.7 inches. Round to the nearest tenth of an inch.

Exercise 5
If the radius of a circle is doubled, how many times larger will its circumference be?

2.4.4 Find the area of a figure.

The area of a figure is a measure of the amount of space contained inside the figure.

The area of a figure is expressed in terms of square units.
For example, if the length and width of a rectangle is measured in inches, the area is expressed in square inches. The same is true for a circle whose radius is measured in inches.

Area Formulas

Rectangle with length L and width W:
$$\text{Area} = \text{Length} \cdot \text{Width} \quad \text{or} \quad A = L \cdot W$$

Square with side x:
$$\text{Area} = \text{Side}^2 \quad \text{or} \quad A = x^2$$

Triangle with base b and height h:
$$\text{Area} = \frac{1}{2}(\text{Base})(\text{Height}) \quad \text{or} \quad A = \frac{1}{2}bh$$

Circle with radius r:
$$\text{Area} = \pi(\text{Radius})^2 \quad \text{or} \quad A = \pi r^2$$

When approximating the area of a circle, use the built-in value of π on your calculator.

Example	Exercise 1
Find the area of a rectangle whose length is 17 inches and whose width is 11 inches. SOLUTION: The formula for the area of a rectangle is $\text{Area} = \text{Length} \cdot \text{Width}$ or $A = L \cdot W$. Substitute 17 for the length and 11 for the width. $$A = L \cdot W$$ $$A = 17 \cdot 11$$ $$A = 187$$ The area is 187 square inches.	Find the area of a rectangle whose width is 16 centimeters and whose length is 1 centimeter more than twice its width.

Exercise 2
Find the area of a triangle whose height is 8 feet and whose base is 3 feet.

Exercise 3
Find the area of a circle whose radius is 8 inches. Round to the nearest tenth of an inch.

Exercise 4
Find the area of a square that has a side of length 8.4 inches.

Exercise 5
A rectangle has a perimeter of 30 inches. If its length is 9 inches, find the area of the rectangle.

2.4.5　Find the volume of a figure.

The volume of a 3-dimensional figure is a measure of the amount of space contained inside the figure. The volume of a figure is expressed in terms of cubic units.

For example, if the length, width, and height of a rectangular solid are measured in inches, the volume is expressed in cubic inches.

Volume Formulas

Rectangular solid with length L, width W, and height H:
$$\text{Volume} = \text{Length} \cdot \text{Width} \cdot \text{Height} \text{ or } V = L \cdot W \cdot H$$

Cube with edge x:
$$\text{Volume} = \text{Edge}^3 \text{ or } V = x^3$$

Right circular cylinder with radius r and height h:
$$\text{Volume} = \pi \left(\text{Radius}\right)^2 \cdot \text{Height} \text{ or } V = \pi r^2 h$$

Volume of a sphere with radius r:
$$\text{Volume} = \frac{4}{3}\pi \left(\text{Radius}\right)^3 \text{ or } V = \frac{4}{3}\pi r^3$$

Volume of a cone with height h and radius r:
$$\text{Volume} = \frac{1}{3}\pi \left(\text{Radius}\right)^2 \cdot \text{Height} \text{ or } V = \frac{1}{3}\pi r^2 h$$

When approximating the volume of a right circular cylinder or a sphere, use the built-in value of π on your calculator.

Example	Exercise 1
Find the volume of a rectangular solid with a length of 6 feet, width of 4 feet, and height of 3 feet. SOLUTION: The formula for the volume of a rectangular solid is $\text{Volume} = \text{Length} \cdot \text{Width} \cdot \text{Height}$ or $V = L \cdot W \cdot H$. Substitute 6 for the length, 4 for the width, and 3 for the height. $$V = L \cdot W \cdot H$$ $$V = 6 \cdot 4 \cdot 3$$ $$V = 72$$ The volume is 72 cubic feet.	Find the volume of a cube whose side is 7 inches.

Exercise 2
The length of a rectangular solid is 3 times its height. The width of the rectangular solid is 4 inches more than its height. If the height is 8 inches, find the volume of the rectangular solid.

Exercise 3
A right circular cylinder has a radius of 4 centimeters and a height of 15 centimeters. Find its volume. Round to the nearest tenth of a cubic centimeter.

Exercise 4
A cone with a height of 8 inches has a circular base with a diameter of 6 inches. Find its volume. Round to the nearest tenth of a cubic inch.

Exercise 5
A ball has a diameter of 9.4 inches. Find its volume. Round to the nearest tenth of a cubic inch.

2.4.6   Solve applications involving distance, rate, and time.

When an object moves at a constant rate of speed r for a time t, the distance traveled, d, is given by the formula $d = r \cdot t$.

The unit for the rate of speed must be the same as the unit for distance over the unit for time. For example, if the distance is measured in miles and the time is measured in hours, then the rate of speed must be in miles/hour.

If the distance is unknown, it can be found by multiplying the rate and the time.

If either the rate or the time are unknown, use the formula $d = r \cdot t$ to set up an equation and use the multiplication property of equality to solve it.

If a problem involves two or more legs of a trip, an equation can often be set up comparing (or combining) the distances of each leg of the trip.

For example, if a driver is making a round trip, then the distance of the two legs must be equal to each other. The distance for each leg is equal to the product of the rate of speed and the time traveled.

If the total distance of the two legs is known, add the distance from the first leg $(\text{rate} \cdot \text{time})$ to the distance of the second leg and set that sum equal to the total distance.

Example	Exercise 1
If Mario drives at a speed of 70 mph, how long will it take him to drive 455 miles? SOLUTION: Substitute 70 for r and 455 for d in the equation $d = r \cdot t$. The equation can then be solved for the unknown time, t. $$d = r \cdot t$$ $$455 = 70t$$ $$\frac{455}{70} = \frac{70t}{70}$$ $$6.5 = t$$ It will take Mario 6.5 hours to drive 455 miles.	A commercial airline flew 3000 miles in 6 hours. What was its average speed.

Exercise 2
Danica drove her race car at an average speed of 125 mph for 4 hours. How far did she drive?

Exercise 3
A bullet train travels 260 kilometers per hour. How far can it travel in 45 minutes?

Exercise 4
Susan averaged 70 mph on the way to Louisville to make a sales call. On the way home, she averaged 60 mph and it took her 1 hour longer to drive home than it took her to drive to Louisville. How long did it take Susan to drive from her home to Louisville?

Exercise 5
Erik drove his truck 220 miles to make a delivery. For part of the trip he had to reduce his speed from an average of 60 mph to 40 mph due to construction. If he spent 2 fewer hours driving at 40 mph than he spent driving at 60 mph, how long was he in the construction zone?

3.1.1 Write ordered pairs.

An ordered pair is used to give the x-coordinate and y-coordinate of a point on a rectangular coordinate plane. Ordered pairs are always written in the form (x, y).

The x-axis is the horizontal axis and the y-axis is the vertical axis.

The point at which the two axes intersect is called the origin, and the ordered pair associated with it is $(0,0)$.

If a point is located to the right of the y-axis then its x-coordinate will be positive.
If a point is located to the left of the y-axis then its x-coordinate will be negative.
If a point is located on the y-axis then its x-coordinate will be 0.

If a point is located above the x-axis then its y-coordinate will be positive.
If a point is located below the x-axis then its y-coordinate will be negative.
If a point is located on the x-axis then its y-coordinate will be 0.

To determine the ordered pair associated with a point on a rectangular coordinate plane, first determine its x-coordinate by finding how far to the right or left of the y-axis it is. Then determine its y-coordinate by finding how far above or below the x-axis it is.

Example	Exercise 1
Identify the ordered pair on the graph.	Identify the ordered pair on the graph.
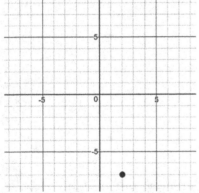	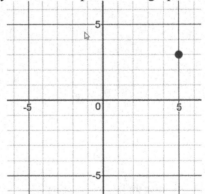
Created using the Desmos Graphing Calculator	Created using the Desmos Graphing Calculator

SOLUTION:
The point is located 2 units to the right of the origin, so its x-coordinate is 2.
The point is located 7 units below the origin, so its y-coordinate is -7.

The ordered pair is $(2, -7)$.

Exercise 2

Identify the ordered pair on the graph.

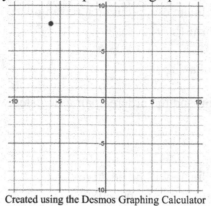

Created using the Desmos Graphing Calculator

Exercise 3

Identify the ordered pair on the graph.

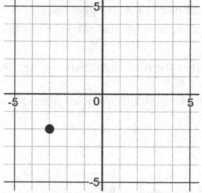

Created using the Desmos Graphing Calculator

Exercise 4

Identify the ordered pair on the graph.

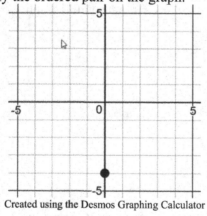

Created using the Desmos Graphing Calculator

Exercise 5

Identify the ordered pair on the graph.

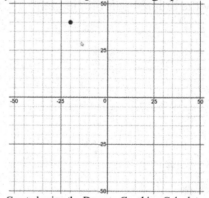

Created using the Desmos Graphing Calculator

3.1.2 Plot points in the rectangular coordinate system.

The first coordinate of an ordered pair tells you how far to the right or left of the y-axis your point will be located.

If the x-coordinate is positive then the point will be to the right of the y-axis, and if it is negative the point will be to the left of the y-axis.

If the x-coordinate is 0 then the point will be located on the y-axis.

The second coordinate of an ordered pair tells you how far above or below the x-axis your point will be located.

If the y-coordinate is positive then the point will be above the x-axis, and if it is negative the point will be below the x-axis.

If the y-coordinate is 0 then the point will be located on the x-axis.

After you plot the point, check to make sure that its coordinates match those of the point you are trying to plot.

Example	Exercise 1
Plot the ordered pair $\left(-4,-5\right)$.	Plot the ordered pair $\left(2,-9\right)$.
SOLUTION: Starting at the origin, move 4 units to the left of the origin. Then move down by 5 units and plot a point there. 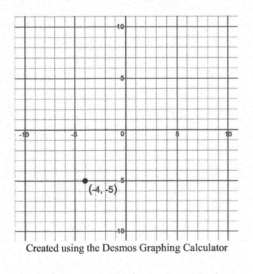 Created using the Desmos Graphing Calculator	

Exercise 2

Plot the ordered pair $(-3,5)$.

Exercise 3

Plot the ordered pair $(0,-4)$.

Exercise 4

Plot the ordered pair $(7,0)$.

Exercise 5

Plot the ordered pair $\left(3\frac{1}{2},-2\frac{1}{4}\right)$.

3.1.3 Complete a table of values of ordered pair solutions for a linear equation in two variables.

To complete a table of ordered pairs, substitute the first given value for the appropriate variable. Then solve for the other variable. Repeat the process for the other values in the table.

Suppose you are asked to complete the following table of ordered pairs for the equation $2x - 5y = -20$.

x	y
-5	
	8
10	

Begin by substituting -5 for x: $2(-5) - 5y = -20$

Then solve the resulting equation for y to complete the ordered pair.

$$-10 - 5y = -20$$
$$-5y = -10$$
$$y = 2$$

The first ordered pair is $(-5, 2)$. Move on to the second ordered pair. Substitute 8 for y:

$2x - 5(8) = -20$. Then solve the resulting equation for x to complete the ordered pair.

$$2x - 40 = -20$$
$$2x = 20$$
$$x = 10$$

The second ordered pair is $(10, 8)$. You would then repeat the process for the third ordered pair.

Example	**Exercise 1**
Compete the table of ordered pairs for the equation $y = 2x + 3$.	Compete the table of ordered pairs for the equation $y = -3x + 6$.

Example

x	y
0	
1	
2	

SOLUTION:
Begin by substituting 0 for x in the equation $y = 2x + 3$.

$$y = 2(0) + 3 = 3$$

Repeat the process for $x = 1$ and $x = 2$.

$$y = 2(1) + 3 = 5$$
$$y = 2(2) + 3 = 7$$

Now complete the table of ordered pairs.

x	y
0	3
1	5
2	7

Exercise 1

x	y
-2	
0	
2	

Exercise 2
Compete the table of ordered pairs for the equation $2x + 3y = 15$.

x	y
-3	
3	
9	

Exercise 3
Compete the table of ordered pairs for the equation $5x - 2y = 10$.

x	y
	0
	10
-8	

Exercise 4
Compete the table of ordered pairs for the equation $y = 4x - 15$.

x	y
$\dfrac{3}{2}$	
0	
$\dfrac{7}{4}$	

Exercise 5
Compete the table of ordered pairs for the equation $3x - 4y = -24$.

x	y
0	
	0
4	

3.1.4 Graph linear equations in two variables using a table of values.

One way to graph a line is to complete a table of ordered pairs. You can then plot the points associated with those ordered pairs, and finish by drawing the straight line that passes through the three points.

Suppose you are asked to graph the equation $y = x - 4$ by completing the following table of ordered pairs.

x	y
0	
1	
2	

Substituting 0 for x produces a value of -4 for y.
Substituting 1 for x produces a value of -3 for y.
Substituting 2 for x produces a value of -2 for y.
Plot the three ordered pairs $(0,-4)$, $(1,-3)$, $(2,-2)$. Then draw the line passing through them.

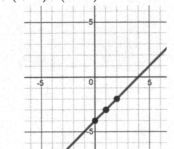

Created using the Desmos Graphing Calculator

Example
Complete the table of ordered pairs and use it to graph $y = 3x - 2$.

x	y
0	
1	
2	

SOLUTION:
Substitute 0, 1, and 2 for x to complete the table of ordered pairs. The three ordered pairs are $(0,-2)$, $(1,1)$, and $(2,4)$. Now plot those three points and draw the line that passes through them.

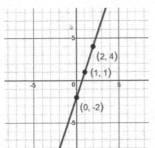

Created using the Desmos Graphing Calculator

Exercise 1
Complete the table of ordered pairs and use it to graph $y = 2x + 5$.

x	y
0	
1	
2	

Exercise 2

Complete the table of ordered pairs and use it to graph $y = -\dfrac{2}{3}x + 8$.

x	y
−3	
0	
3	

Exercise 3

Complete the table of ordered pairs and use it to graph $3x + 2y = 12$.

x	y
0	
	0
2	

Exercise 4

Complete the table of ordered pairs and use it to graph $5x - 4y = 40$.

x	y
	−10
	−5
	0

Exercise 5

Complete the table of ordered pairs and use it to graph $-3x + 4y = 18$.

x	y
0	
	3
2	

3.2.1 Find the intercepts of a line.

The x-intercept of a line is a point where the line intersects the x-axis.

Since all points on the x-axis have a y-coordinate of 0, the x-intercept can be found by substituting 0 for y and solving for x.

The x-intercept will always be of the form $(a, 0)$.

The y-intercept of a line is a point where the line intersects the y-axis.

Since all points on the y-axis have an x-coordinate of 0, the y-intercept can be found by substituting 0 for x and solving for y.

The y-intercept will always be of the form $(0, b)$.

Some lines will only have one intercept.
- Horizontal lines, other than $y = 0$, will only have a y-intercept.
- Vertical lines, other than $x = 0$, will only have an x-intercept.
- Some lines will have their intercepts at the origin. In this case the x-intercept and the y-intercept are the same point.

Example	Exercise 1
Find the x- and y-intercepts of the line $2x + y = 6$.	Find the x- and y-intercepts of the line $4x - 3y = -24$.

Example (continued):

SOLUTION:
To find the x-intercept, substitute 0 for y and solve for x.

$$2x + (0) = 6$$
$$2x = 6$$
$$x = 3$$

The x-intercept is $(3, 0)$.

To find the y-intercept, substitute 0 for x and solve for y.

$$2(0) + y = 6$$
$$0 + y = 6$$
$$y = 6$$

The y-intercept is $(0, 6)$.

Exercise 2
Find the x- and y-intercepts of the line $5x + 2y = -10$.

Exercise 3
Find the x- and y-intercepts of the line $x - 4y = 8$.

Exercise 4
Find the x- and y-intercepts of the line $4x + 3y = -18$.

Exercise 5
Find the x- and y-intercepts of the line $25x + 30y = 1050$.

3.2.2 Graph a linear equation in two variables given its intercepts.

One strategy for graphing a line is to find its intercepts. Once you have found the intercepts you can draw the straight line passing through the two points.

The x-intercept can be found by substituting 0 for y and solving for x.

The y-intercept can be found by substituting 0 for x and solving for y.

- Find the x-intercept and plot it on the graph.
- Find the y-intercept and plot it on the graph.
- Draw the straight line passing through the two intercepts.

Finding the coordinates of a third point can serve as a way to check your work when finding the intercepts. If the third point does not lie on the same straight line that passes through the intercepts then you should go back and check your work for all three points.

Some lines will only have one intercept and will require an alternate strategy.
- Horizontal lines: Find the y-intercept and draw the horizontal line passing through it.
- Vertical lines: Find the x-intercept and draw the vertical line passing through it.
- Lines passing through the origin: Find a second point on the line by selecting a value for x and finding the corresponding value of y.

Example	Exercise 1
Graph $6x+4y=-36$ using its intercepts.	Graph $x+3y=6$ using its intercepts.

SOLUTION:
Begin by finding the intercepts.

$$x-\text{intercept} \qquad y-\text{intercept}$$
$$6x+4(0)=-36 \quad 6(0)+4y=-36$$
$$6x=-36 \qquad 4y=-36$$
$$x=-6 \qquad y=-9$$
$$(-6,0) \qquad (0,-9)$$

Plot the points and draw the line passing through them.

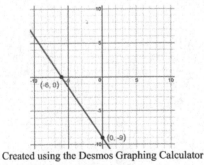

Created using the Desmos Graphing Calculator

Exercise 2
Graph $5x - 2y = 10$ using its intercepts.

Exercise 3
Graph $-3x + 2y = -18$ using its intercepts.

Exercise 4
Graph $y = 2x - 4$ using its intercepts.

Exercise 5
Graph $y = -3x + 9$ using its intercepts.

3.3.1 Find the slope of a line using the slope formula.

Slope Formula

If a line passes through two points (x_1, y_1) and (x_2, y_2), we can calculate its slope using the formula

$$m = \frac{y_2 - y_1}{x_2 - x_1}$$

Slope is the ratio of vertical change (rise) to horizontal change (run).

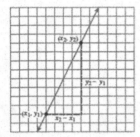

The numerator, $y_2 - y_1$, represents the vertical change.

The denominator, $x_2 - x_1$, represents the horizontal change.

When computing the slope of a line passing through two points, simplify the numerator and denominator separately, then simplify the fraction to simplest terms. If the result is an improper fraction, leave it in that form instead of converting it to a mixed number. This leaves the slope in a form comparing the vertical change to the horizontal change.

- For a horizontal line the vertical change is 0. Thus, the slope of a horizontal line is 0.
- For a vertical line the horizontal change is 0. Thus, the slope of a vertical line is undefined.

Example	Exercise 1
Find the slope of the line passing through $(-2,3)$ and $(1,12)$.	Find the slope of the line passing through $(3,18)$ and $(5,8)$.
SOLUTION: Start by labeling $(-2,3)$ as (x_1, y_1) and $(1,12)$ as (x_2, y_2), then use the slope formula. $$m = \frac{y_2 - y_1}{x_2 - x_1}$$ $$m = \frac{12-3}{1-(-2)} = \frac{12-3}{1+2} = \frac{9}{3} = 3$$	

Exercise 2
Find the slope of the line passing through $(-6,-7)$ and $(2,15)$.

Exercise 3
Find the slope of the line passing through $(-9,18)$ and $(3,-12)$.

Exercise 4
Find the slope of the line passing through $(-3,8)$ and $(10,8)$.

Exercise 5
Find the slope of the line passing through $(5,4)$ and $(5,-6)$.

3.3.2   Find the slope of a line given its graph.

Slope is the ratio of vertical change (rise) to horizontal change (run).

To determine the slope of a line from its graph, begin by identifying two points on the graph. Preferably, those two points will have integer coordinates.

Determine the vertical change and horizontal change of the line between those two points. This can be done by drawing a triangle as shown below.

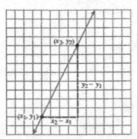

The quantity $y_2 - y_1$ represents the vertical change (rise).

The quantity $x_2 - x_1$ represents the horizontal change (run).

Express the slope as a fraction $\left(\dfrac{\text{rise}}{\text{run}} \right)$ and simplify the fraction.

Example	**Exercise 1**
Find the slope of the line.	Find the slope of the line.
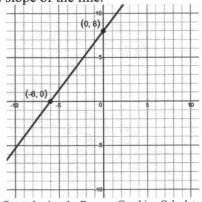	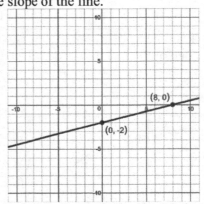
Created using the Desmos Graphing Calculator	Created using the Desmos Graphing Calculator

SOLUTION:

Two points on this line whose coordinates we know are the x-intercept $(-6,0)$ and the y-intercept $(0,8)$. Starting at the point $(-6,0)$, count up 8 units to reach the same level as $(0,8)$. This is the rise. Count 6 units to the left, which is the run.

$$m = \frac{\text{rise}}{\text{run}} = \frac{8}{6} = \frac{4}{3}$$

So, the slope is $\dfrac{4}{3}$.

Exercise 2
Find the slope of the line.

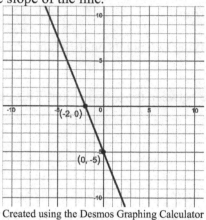

Created using the Desmos Graphing Calculator

Exercise 3
Find the slope of the line.

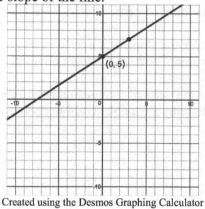

Created using the Desmos Graphing Calculator

Exercise 4
Find the slope of the line.

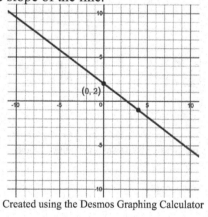

Created using the Desmos Graphing Calculator

Exercise 5
Find the slope of the line.

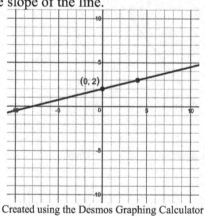

Created using the Desmos Graphing Calculator

3.3.3 Graph a line given its equation in slope-intercept form.

Slope-Intercept Form
If a line has slope m and y-intercept $(0,b)$, the slope-intercept form of the line is $y = mx + b$.

To graph a line whose equation is in slope-intercept form, $y = mx + b$, first plot a point at the y-intercept $(0,b)$.

Next use the slope m to find a second point on the line and plot it. If the slope is expressed as a fraction, the numerator tells you how many units above or below the y-intercept to move. The denominator tells you how many units to move to the right of the y-intercept.

- For example, if $m = -\dfrac{2}{5}$, move down 2 units (because the numerator is negative) and 5 units to the right of the y-intercept.

- If $m = 4$, we can think of this as $m = \dfrac{4}{1}$. Move up 4 units (because the numerator is positive) and 1 unit to the right.

Finish by graphing the line that passes through those two points.

Example	**Exercise 1**
Graph $y = -\dfrac{3}{4}x + 2$ using its slope and y-intercept.	Graph $y = 5x - 4$ using its slope and y-intercept.
SOLUTION: Plot the y-intercept at $(0,2)$ and use the slope of to find a second point (down 3, right 4). 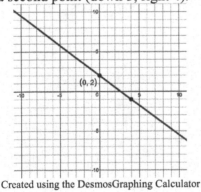 Created using the DesmosGraphing Calculator	

Exercise 2

Graph $y = -x - 7$ using its slope and y-intercept.

Exercise 3

Graph $y = \frac{2}{3}x + 5$ using its slope and y-intercept.

Exercise 4

Graph $y = -\frac{4}{7}x + 5$ using its slope and y-intercept.

Exercise 5

Graph $y = 4x$ using its slope and y-intercept.

3.3.4 Graph a line given one point on the line and the slope.

Suppose the slope m of a line is known, as well as the coordinates of a point (h,k) on the line. Graphing a line under these conditions is similar to graphing a line in slope-intercept form.

First plot a point at (h,k).

Then use the slope m to find a second point on the line.

- For example, if $m = \dfrac{3}{2}$, move up 3 units (because the numerator is positive) and 2 units to the right of the point (h,k).

- If $m = -3$, we can think of this as $m = \dfrac{-3}{1}$. Move down 3 units (because the numerator is negative) and 1 unit to the right.

Finish by graphing the line that passes through those two points.

The equation of a line with these conditions can be expressed as $y = m(x-h) + k$.

Example	Exercise 1
Graph the line that passes through the point $(2,3)$ with a slope of 4.	Graph the line that passes through the point $(5,-1)$ with a slope of -2.
SOLUTION: Start by plotting a point at $(2,3)$. Next use the slope of 4 to find a second point on the line, which is 4 units above and 1 unit to the right of $(2,3)$. Graph the line that passes through the two points. 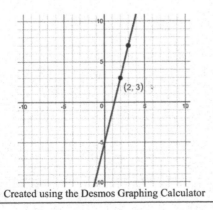 Created using the Desmos Graphing Calculator	

Exercise 2

Graph the line that passes through the point $(-6, -2)$ with a slope of $\dfrac{3}{4}$.

Exercise 3

Graph the line that passes through the point $(-4, 0)$ with a slope of $-\dfrac{2}{5}$.

Exercise 4

Graph the line that passes through the point $(-1, -3)$ with a slope of 6.

Exercise 5

Graph the line that passes through the point $(-4, 5)$ with a slope of $\dfrac{1}{7}$.

3.3.5   Graph vertical lines.

Vertical Lines

A vertical line has an equation of the form $x = a$.

Note that the equation does not contain the variable y, only the variable x.

For an equation of the form $x = a$, all points on the line have an x-coordinate of a. Looking at all the points that have the same x-coordinate, we see that they all lie on a vertical line.

To graph a vertical line $x = a$,

- Plot a point at the x-intercept $(a, 0)$.
- Graph the vertical line passing through that point.

Example	**Exercise 1**
Graph the line $x = 5$.	Graph the line $x = -2$.

Example

Graph the line $x = 5$.

SOLUTION:

Start by plotting the x-intercept at $(5, 0)$.

Then graph the vertical line through that point.

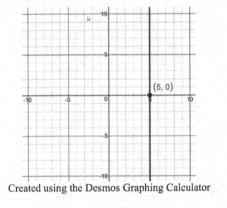

Created using the Desmos Graphing Calculator

Exercise 1

Graph the line $x = -2$.

Exercise 2
Graph the line $x = -8$.

Exercise 3
Graph the line $x = 3$.

Exercise 4
Graph the line $x = -3\dfrac{1}{2}$.

Exercise 5
Graph the line $x = 0$.

3.3.6 Graph horizontal lines.

Horizontal Lines
A horizontal line has an equation of the form $y = b$.

Note that the equation does not contain the variable x, only the variable y.

For an equation of the form $y = b$, all points on the line have a y-coordinate of b. Looking at all the points that have the same b-coordinate, we see that they all lie on a horizontal line.

To graph a horizontal line $y = b$,

- Plot a point at the y-intercept $(0, b)$.
- Graph the horizontal line passing through that point.

Example	Exercise 1
Graph the line $y = 5$.	Graph the line $y = -9$.
SOLUTION:	
Start by plotting the y-intercept at $(0, 5)$.	
Then graph the horizontal line through that point.	

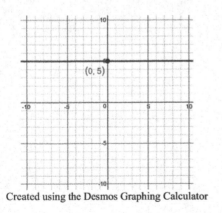

Created using the Desmos Graphing Calculator

Exercise 2
Graph the line $y = -1$.

Exercise 3
Graph the line $y = 6$.

Exercise 4
Graph the line $y = -\dfrac{3}{4}$.

Exercise 5
Graph the line $y = 0$.

3.3.7   Use slope with parallel and perpendicular lines.

Parallel Lines
- Two lines that do not intersect are parallel lines.
- Two nonvertical lines are parallel if they have the same slope.

  For example, if a line has slope $m = \dfrac{2}{3}$, a parallel line would have to also have slope $m = \dfrac{2}{3}$.
- A vertical line is parallel to another vertical line.

Perpendicular Lines
- Two lines that intersect at a 90° angle are perpendicular lines.
- Two nonvertical lines are perpendicular if their slopes are negative reciprocals.

  In other words, if the slope of the first line is m_1 then the slope of a perpendicular line is

  $-\dfrac{1}{m_1}$. For example, if a line has slope $m = \dfrac{2}{3}$, a perpendicular line would have to have slope

  $m = -\dfrac{3}{2}$.
- A vertical line is perpendicular to a horizontal line.

To determine if two nonvertical lines are parallel or perpendicular, begin by finding the slope of each line. This is done by writing the equation of each line in slope-intercept form $(y = mx + b)$.

If the two slopes are equal then the lines are parallel.

If the two slopes have opposite signs and are reciprocals then the lines are perpendicular.

Example	Exercise 1
Are the lines $y = -\dfrac{3}{4}x + 2$ and $6x + 8y = -40$ parallel, perpendicular, or neither? SOLUTION: The equation $y = -\dfrac{3}{4}x + 2$ is in slope-intercept form, and its slope is $m = -\dfrac{3}{4}$. To find the slope of $6x + 8y = -40$, solve for y. $6x + 8y = -40 \;\rightarrow\; y = -\dfrac{3}{4}x - 5$ The second line also has a slope of $m = -\dfrac{3}{4}$, so the two lines are parallel.	Are the lines $y = 2x + 7$ and $10x - 5y = 20$ parallel, perpendicular, or neither?

Exercise 2

Are the lines $y = -\dfrac{2}{7}x + 8$ and $6x + 21y = -42$ parallel, perpendicular, or neither?

Exercise 3

Are the lines $3x + 4y = 36$ and $20x + 15y = -60$ parallel, perpendicular, or neither?

Exercise 4

Are the lines $4x + 6y = 60$ and $9x - 6y = 30$ parallel, perpendicular, or neither?

Exercise 5

Are the lines $3x + 5y = 10$ and $10x + 6y = -30$ parallel, perpendicular, or neither?

3.3.8 Interpret slope as a rate of change.

The slope of a line describes the rate of change of y with respect to x.

The vertical change (rise) is the change in y as x changes by an amount equal to the horizontal change (run).

If the slope of the line is m, we can say that the rate of change of y is m.
y changes by m units as x increases by 1 unit.
- If the slope is positive, then the rate of change is also positive and y increases by m units for each unit that x increases.
- If the slope is negative, then the rate of change is also negative and y decreases by m units for each unit that x increases.

If the slope is not an integer, the rate of change can be expressed as a non-unit rate.
The numerator of the slope represents the rate of change of y when x increases by the number of units in the denominator.

For example, if $m = \dfrac{2}{5}$, then we can say that y increases by 2 units as x increases by 5 units.

Example	**Exercise 1**
Interpret the slope of $y = -2x + 9$ as a rate of change.	Interpret the slope of $y = x + 6$ as a rate of change.
SOLUTION: The slope is -2, which can be expressed as a fraction as $\dfrac{-2}{1}$. This indicates that y decreases by 2 when x increases by 1.	

Exercise 2 Interpret the slope of $y = \frac{3}{5}x - 1$ as a rate of change.	**Exercise 3** Interpret the slope of $y = -\frac{4}{7}x - 8$ as a rate of change.
Exercise 4 Interpret the slope of $y = 0.2x - 15$ as a rate of change.	**Exercise 5** Interpret the slope of $y = -3$ as a rate of change.

3.4.1 Find the slope of a line given its equation.

The slope-intercept form of a line is $y = mx + b$.

If a linear equation in two variables is rewritten so that y is isolated on one side the equation, then the slope of the equation is the coefficient of the term containing x. (The y-coordinate of the y-intercept is the constant term.)

Slope-Intercept Form of a Line
- $y = mx + b$
- Slope: m
- y-intercept: $(0, b)$

To find the slope of a line from its equation, first rewrite it in slope-intercept form by solving the equation for y in terms of x.
This is similar to solving a formula for a particular variable, or solving a literal equation.

Use the addition property of equality to isolate the term containing y on one side of the equation.
Solve the equation by dividing both sides of the equation by the coefficient of the term containing y.
Simplify the coefficient of the term containing x to find the slope of the line.
(Simplify the constant term to find the y-coefficient of the y-intercept.)

Example	**Exercise 1**
Find the slope of the line $6x - 4y = 12$.	Find the slope of the line $y = x - 9$.
SOLUTION: To find the slope, begin by rewriting the equation in slope-intercept form. $(y = mx + b)$ $$6x - 4y = 12$$ $$-4y = -6x + 12$$ $$\frac{-4y}{-4} = \frac{-6x}{-4} + \frac{12}{-4}$$ $$y = \frac{3}{2}x - 3$$ The slope of the line is $m = \frac{3}{2}$.	

Exercise 2

Find the slope of the line $y = -\dfrac{6}{5}x + 7$.

Exercise 3

Find the slope of the line $y = \dfrac{2}{9}$.

Exercise 4

Find the slope of the line $8x + 6y = -30$.

Exercise 5

Find the slope of the line $-12x + 8y = -72$.

3.4.2   Write the slope-intercept form of a line.

The slope-intercept form of a line is $y = mx + b$.

If the slope of the line and its y-intercept are known, then the equation of the line can be written in slope-intercept form of a line.

- The slope of the line is m.
- The y-coordinate of the y-intercept is b.
- Write the equation in slope-intercept form, $y = mx + b$, by substituting the appropriate values for m and b.

If a line has a slope of 0, then the line is a horizontal line.
Its equation will be of the form $y = b$, where b is the y-coordinate of the y-intercept.

Example	**Exercise 1**
Write the slope-intercept form of the equation of a line whose slope is -2 and whose y-intercept is $(0,4)$.	Write the slope-intercept form of the equation of a line whose slope is 5 and whose y-intercept is $(0,8)$.
SOLUTION: To write the equation in slope-intercept form, find the slope (m) and the y-coordinate of the y-intercept (b). Then substitute those values for m and b in the equation $y = mx + b$. The slope is -2, so $m = -2$. The y-intercept is $(0,4)$, so $b = 4$. The equation of the line is $y = -2x + 4$.	

Exercise 2
Write the slope-intercept form of the equation of a line whose slope is $-\dfrac{4}{3}$ and whose y-intercept is $(0,-6)$.

Exercise 3
Write the slope-intercept form of the equation of a line whose slope is $\dfrac{2}{9}$ and whose y-intercept is $-\dfrac{5}{3}$.

Exercise 4
Write the slope-intercept form of the equation of a line whose slope is 1 and whose y-intercept is $(0,0)$.

Exercise 5
Write the equation of a line whose slope is 0 and whose y-intercept is $(0,16)$.

3.4.3   Write the equation of a line given the slope and a point on the line.

The equation of a line can be found if the slope of the line is known, as well as the coordinates of a point on the line.

One way to find the equation is to use the slope-intercept form of a line, $y = mx + b$.

- Substitute the slope of the line for m.
- Substitute the coordinates of the point on the line for x and y.
- Solve the equation for b.
- Now that m and b are known, write the equation in $y = mx + b$ form.

Another way to find the equation is to use the point-slope form of a line, $y - y_1 = m(x - x_1)$.

- Substitute the slope of the line for m.
- Substitute the coordinates of the point on the line for x_1 and y_1.
- Solve the equation for y.
- The equation of the line is now in $y = mx + b$ form.

Example	Exercise 1
Find the slope-intercept equation of a line with slope -4 that passes through the point $(5, -8)$.	Find the slope-intercept equation of a line with slope 8 that passes through the point $(2, 17)$.
SOLUTION: Since the slope is -4, the equation is of the form $y = -4x + b$. The line passes through the point $(5, -8)$. Substitute 5 for x and -8 for y in the equation $y = -4x + b$, and solve for b. $$-8 = -4(5) + b$$ $$-8 = -20 + b$$ $$12 = b$$ Since $b = 12$, the equation is $y = -4x + 12$.	

Exercise 2
Find the slope-intercept equation of a line with slope 5 that passes through the point $(-7,-9)$.

Exercise 3
Find the slope-intercept equation of a line with slope $-\dfrac{2}{3}$ that passes through the point $(-12,20)$.

Exercise 4
Find the slope-intercept equation of a line with slope $\dfrac{3}{4}$ that passes through the point $(20,51)$.

Exercise 5
Find the slope-intercept equation of a line with slope 18 that passes through the point $\left(\dfrac{7}{6},15\right)$.

3.4.4 Write the equation of a line through two given points.

The equation of a line can be found if the slope of the line is known, as well as the coordinates of a point on the line.

If only the coordinates of two points are known, the slope can be computed using the slope formula:

$$m = \frac{y_2 - y_1}{x_2 - x_1}.$$

The coordinates of either point can be used along with the slope to find the equation of the line.

If the slope of the line is 0 then the line is a horizontal line and its equation is of the form $y = b$.

If the slope of the line is undefined then the line is a vertical line and its equation is of the form $x = a$.

One way to find the equation is to use the slope-intercept form of a line, $y = mx + b$.

- Substitute the slope of the line for m.
- Substitute the coordinates of one of the points on the line for x and y.
- Solve the equation for b.
- Now that m and b are known, write the equation in $y = mx + b$ form.

Another way to find the equation is to use the point-slope form of a line, $y - y_1 = m(x - x_1)$.

- Substitute the slope of the line for m.
- Substitute the coordinates of one of the points on the line for x_1 and y_1.
- Solve the equation for y.
- The equation of the line is now in $y = mx + b$ form.

Example	**Exercise 1**
Find the slope-intercept equation of a line that passes through the points $(2,9)$ and $(7,4)$.	Find the slope-intercept equation of a line that passes through the points $(-2,-5)$ and $(-5,7)$.
SOLUTION: Begin by computing the slope. $$m = \frac{4-9}{7-2} = \frac{-5}{5} = -1$$ Use one of the points to find the value of b. Using the point $(2,9)$, substitute -1 for m, 2 for x and 9 for y in the equation $y = mx + b$, and solve for b. $$y = mx + b$$ $$9 = -1(2) + b$$ $$9 = -2 + b$$ $$11 = b$$ Since $m = -1$ and $b = 11$, the equation is $y = -x + 11$.	

Exercise 2
Find the slope-intercept equation of a line that passes through the points $(-8, 12)$ and $(-3, -3)$.

Exercise 3
Find the slope-intercept equation of a line that passes through the points $(-6, 8)$ and $(15, -20)$.

Exercise 4
Find the equation of a line that passes through the points $(-2, -9)$ and $(-2, -3)$.

Exercise 5
Find the equation of a line that passes through the points $(-4, 8)$ and $(2, 8)$.

4.1.1 Evaluate exponential expressions with positive exponents.

An expression of the form b^n is an exponential expression.

b is the base, and n is the exponent.

This is shorthand notation for repeated multiplication. The base b is written as a factor n times.

For example, for the expression 9^4, the base is 9 and the exponent is 4. So, 9 is a factor 4 times.

$$9^4 = 9 \cdot 9 \cdot 9 \cdot 9$$
$$= 6561$$

Although the notation is used to represent repeated multiplication, you should become familiar with the built-in function on your calculator for raising a base to a given power.

The use of parentheses is necessary for bases that are negative. The base is the number or expression that the exponent is written directly after.

For the expression $(-7)^2$, the base is -7. So, $(-7)^2 = (-7)(-7) = 49$.

However, for the expression -7^2, the base is 7. That means that $-7^2 = -7 \cdot 7$ or -49.

Example	Exercise 1
Simplify 2^5.	Simplify 5^2.
SOLUTION:	
The base is 2 and the exponent is 5, so write 2 as a factor 5 times.	
$$2^5 = 2 \cdot 2 \cdot 2 \cdot 2 \cdot 2$$ $$= 32$$	

Exercise 2
Simplify -3^4.

Exercise 3
Simplify $(-3)^4$.

Exercise 4
Simplify $\left(\dfrac{2}{3}\right)^3$.

Exercise 5
For what value of x is 4^x equal to 1024?

4.1.2 Use the product rule for exponents.

Product Rule for Exponents
For any base x, $x^m \cdot x^n = x^{m+n}$.

This product rule states that when multiplying two expressions with the same base, the product will have the same base and the exponent will be equal to the sum of the exponents from the two expressions.

To gain an understanding of this rule, consider the expression $x^4 \cdot x^3$.
The expression x^4 is equivalent to the base x being used as a factor 4 times: $x^4 = x \cdot x \cdot x \cdot x$.
Similarly, the expression x^3 is equivalent to the base x being used as a factor 3 times: $x^3 = x \cdot x \cdot x$.
So, $x^4 \cdot x^3$ is equal to $x \cdot x \cdot x \cdot x \cdot x \cdot x \cdot x$ and the base x is listed as a factor 7 times.
This is consistent with the rule: $x^4 \cdot x^3 = x^{4+3}$.

The product rule also applies when 3 or more exponential expressions are being multiplied.

Keep in mind that the product rule only applies to exponents, not coefficients.
To simplify the expression $9x^4 \cdot 5x^3$, begin by multiplying the coefficients 9 and 5. Then add the exponents for x.

$$9x^4 \cdot 5x^3 = (9 \cdot 5) x^{4+3}$$
$$= 45x^7$$

Example	**Exercise 1**
Simplify $x^9 \cdot x^4$.	Simplify $x^{13} \cdot x^{47}$.
SOLUTION: Apply the product rule for exponents by adding the two exponents. $$x^9 \cdot x^4 = x^{9+4}$$ $$= x^{13}$$	

Exercise 2

Simplify $x \cdot x^{12} \cdot x^6$.

Exercise 3

Simplify $(x+8)^6 \cdot (x+8)^4$.

Exercise 4

Simplify $x^{14} \cdot x^7 \cdot x^{29}$.

Exercise 5

Simplify $8x^9 y^5 \cdot 5x^2 y$.

4.1.3 Use the power rules for exponents.

The power rule for exponents is used when raising an exponential expression to a power.
Power Rule for Exponents

$$\text{For any base } x, \ \left(x^m\right)^n = x^{m \cdot n}.$$

This product rule states that when raising an exponential expression to a power, keep the base and multiply the exponents.

To gain an understanding of this rule, consider the expression $\left(x^4\right)^3$.

List x^4 as a factor 3 times: $\left(x^4\right)^3 = x^4 \cdot x^4 \cdot x^4$.

By the product rule, this is equivalent to x^{4+4+4} or $x^{4 \cdot 3}$.

The power of a product rule for exponents is used when raising a product to a power.
Power of a Product Rule for Exponents

$$\text{For any bases } x \text{ and } y, \ \left(xy\right)^n = x^n y^n.$$

This rule states that when raising a product to a power, raise each factor to that power.

The expression $\left(xy\right)^4$ is equivalent to $xy \cdot xy \cdot xy \cdot xy$. Using the commutative property to rearrange the factors, this is equivalent to $x \cdot x \cdot x \cdot x \cdot y \cdot y \cdot y \cdot y$. This can be rewritten as $x^4 y^4$.

Example	**Exercise 1**
Simplify $\left(x^6\right)^7$.	Simplify $\left(x^9\right)^2$.
SOLUTION: Apply the power rule for exponents by multiplying the two exponents. $$\left(x^6\right)^7 = x^{6 \cdot 7}$$ $$= x^{42}$$	

Exercise 2

Simplify $\left(x^{13}\right)^{14}$.

Exercise 3

Simplify $\left(x^2\right)^6 \cdot \left(x^{10}\right)^4$.

Exercise 4

Simplify $\left(x^8 y^{18}\right)^4$.

Exercise 5

Simplify $\left(9x^3 y^{11} z^6\right)^3$.

4.1.4 Use the quotient rule for exponents.

The quotient rule for exponents is used with fractions that have the same base in the numerator and denominator.

Quotient Rule for Exponents

$$\text{For any base } x, \frac{x^m}{x^n} = x^{m-n} \ \left(x \neq 0 \right).$$

This quotient rule states that when the same base appears in the numerator and denominator of a fraction, keep the base and subtract the exponents.

To gain an understanding of this rule, consider the expression $\frac{x^9}{x^3}$.

First, rewrite x^9 and x^3.

$$\frac{x^9}{x^3} = \frac{x \cdot x \cdot x \cdot x \cdot x \cdot x \cdot x \cdot x \cdot x}{x \cdot x \cdot x}$$

Now divide out the common factors $\left(x \cdot x \cdot x \right)$ in the numerator and denominator.

$$\frac{\overset{1}{\cancel{x}} \cdot \overset{1}{\cancel{x}} \cdot \overset{1}{\cancel{x}} \cdot x \cdot x \cdot x \cdot x \cdot x \cdot x}{\underset{1}{\cancel{x}} \cdot \underset{1}{\cancel{x}} \cdot \underset{1}{\cancel{x}}}$$

The resulting expression is equivalent to x^6, which can be found by subtracting the exponents: x^{9-3}.

Example	Exercise 1
Simplify $\dfrac{x^{20}}{x^4}$. Assume all variables are nonzero.	Simplify $\dfrac{x^7}{x^3}$. Assume all variables are nonzero.
SOLUTION: Apply the quotient rule for exponents by subtracting the two exponents. $$\frac{x^{20}}{x^4} = x^{20-4}$$ $$= x^{16}$$	

Exercise 2

Simplify $\dfrac{x^{10}}{x^9}$. Assume all variables are nonzero.

Exercise 3

Simplify $\dfrac{x^{16}}{x}$. Assume all variables are nonzero.

Exercise 4

Simplify $\dfrac{x^6 y^{15}}{x^3 y^{10}}$. Assume all variables are nonzero.

Exercise 5

Simplify $\dfrac{16x^{16}}{8x^8}$. Assume all variables are nonzero.

4.1.5   Evaluate exponential expressions with integer exponents.

Negative Exponents

$$\text{For any nonzero base } x, \quad x^{-n} = \frac{1}{x^n}.$$

For example, 7^{-2} is equivalent to $\frac{1}{7^2}$ or $\frac{1}{49}$.

To gain an understanding of this rule, consider the expression $\frac{x^3}{x^8}$.

By the quotient rule, $\frac{x^3}{x^8} = x^{3-8} = x^{-5}$.

Dividing out common factors: $\frac{x^3}{x^8} = \frac{x \cdot x \cdot x}{x \cdot x \cdot x \cdot x \cdot x \cdot x \cdot x \cdot x} = \frac{\overset{1}{\cancel{x}} \cdot \overset{1}{\cancel{x}} \cdot \overset{1}{\cancel{x}}}{\underset{1}{\cancel{x}} \cdot \underset{1}{\cancel{x}} \cdot \underset{1}{\cancel{x}} \cdot x \cdot x \cdot x \cdot x \cdot x} = \frac{1}{x^5}$.

Since both x^{-5} and $\frac{1}{x^5}$ are both equal to $\frac{x^3}{x^8}$, they are equal to each other by the transitive property.

If an expression has a base with a negative exponent, the first step is to rewrite the expression so that all exponents are nonnegative. Then simplify if possible.

Example	**Exercise 1**
Simplify 2^{-3}.	Simplify 5^{-2}.
SOLUTION: First rewrite the expression without negative exponents. $$2^{-3} = \frac{1}{2^3}$$ Now simplify 2^3, which is equal to 8. $$2^{-3} = \frac{1}{2^3} = \frac{1}{8}$$	

Exercise 2
Simplify 10^{-5}.

Exercise 3
Simplify -9^3.

Exercise 4
Simplify 9^{-3}.

Exercise 5
Simplify $\left(\dfrac{3}{5}\right)^{-4}$.

4.1.6 Simplify exponential expressions using the rules for exponents.

Here is a summary of the exponent rules.

Exponent Rules

1. Product Rule

   For any base x, $x^m \cdot x^n = x^{m+n}$.

2. Power Rule

   For any base x, $\left(x^m\right)^n = x^{m \cdot n}$.

3. Power of a Product Rule

   For any bases x and y, $(xy)^n = x^n y^n$.

4. Quotient Rule

   For any base x, $\dfrac{x^m}{x^n} = x^{m-n}$ $(x \neq 0)$.

5. Zero Exponent Rule

   For any base x, $x^0 = 1$ $(x \neq 0)$.

6. Power of a Quotient Rule

   For any bases x and y, $\left(\dfrac{x}{y}\right)^n = \dfrac{x^n}{y^n}$ $(y \neq 0)$.

7. Negative Exponent Rule

   For any base x, $x^{-n} = \dfrac{1}{x^n}$ $(x \neq 0)$.

Example	Exercise 1
Simplify the expression $x^{-17} \cdot x^8$. Write the result without negative exponents. (Assume that $x \neq 0$.) SOLUTION: Begin by applying the product rule for exponents. $$x^{-17} \cdot x^8 = x^{-17+8}$$ $$= x^{-9}$$ Now rewrite the expression without negative exponents. $$x^{-17} \cdot x^8 = x^{-17+8}$$ $$= x^{-9}$$ $$= \frac{1}{x^9}$$	Simplify the expression $\left(x^7\right)^6$.

Exercise 2

Simplify the expression $\dfrac{x^5}{x^{18}}$. Write the result without negative exponents. (Assume that $x \neq 0$.)

Exercise 3

Simplify the expression $\left(2x^3 y^4\right)^6$.

Exercise 4

Simplify the expression $\left(7x^{-3} y^2 z^{-8}\right)^{-2}$. Write the result without negative exponents. (Assume that all variables are not equal to 0.)

Exercise 5

Simplify the expression $\left(\dfrac{x^{-3} y^2}{z^{-5}}\right)^{-8}$. Write the result without negative exponents. (Assume that all variables are not equal to 0.)

4.2.1 Convert between scientific and standard notation.

Scientific notation is used to represent numbers that are very large, such as 93,000,000, or very small, such as 0.0000324.

A number in scientific notation is of the form $a \times 10^b$, where $1 \leq a < 10$ and b is an integer.
- Positive powers of 10 are equal to numbers that are 10 or higher.
- Negative powers of 10 are equal to numbers that are between 0 and 1.

To convert a number to scientific notation, first move the decimal point so that it immediately follows the first nonzero digit in the number.
Count the number of decimal places the decimal point moves. This gives the power of 10 when the number is written in scientific notation.
If the original number is 10 or larger, the exponent is positive.
If the original number is between 0 and 1, the exponent is negative.

To rewrite a number that is in scientific notation $\left(a \times 10^b \right)$ as a number in standard notation, multiply the decimal number a by 10^b.

If b is positive, multiplying by 10^b is equivalent to moving the decimal point in a by b places to the right.
If b is negative, multiplying by 10^b is equivalent to moving the decimal point in a by b places to the left.

Example	Exercise 1
Rewrite 2,400,000 in scientific notation. SOLUTION: Place the decimal point after the first nonzero digit, 2. Since the decimal point will be moved 6 places to the left, the power of 10 is 6. $$2,400,000 = 2.4 \times 10^6$$	Rewrite 0.00135 in scientific notation.

Exercise 2
Rewrite 49,200 in scientific notation.

Exercise 3
Rewrite 7.29×10^{-7} in standard notation.

Exercise 4
Rewrite 4.8×10^{5} in standard notation.

Exercise 5
Rewrite 1.004×10^{9} in standard notation.

4.2.2 Perform calculations involving scientific notation.

To multiply two numbers that are in scientific notation, multiply the two decimal numbers by each other. Then use the product property of exponents $\left(x^m \cdot x^n = x^{m+n}\right)$ to multiply the powers of 10. This strategy comes from the commutative and associative properties of arithmetic.

$$\left(a \times 10^b\right) \cdot \left(c \times 10^d\right) = a \times 10^b \cdot c \cdot 10^d$$
$$= a \cdot c \times 10^b \times 10^d$$
$$= (ac) \times 10^{b+d}$$

To divide two numbers that are in scientific notation, divide the first decimal number by the second decimal number. Then use the quotient property of exponents $\left(x^m \div x^n = x^{m-n}\right)$ to divide the powers of 10.

$$\left(a \times 10^b\right) \div \left(c \times 10^d\right) = \frac{a}{c} \times 10^{b-d}$$

After multiplying or dividing, be sure that the number is in scientific notation.
If the decimal number is less than 1, move the decimal point to the right by the appropriate number of places, and subtract that number from the exponent of 10.
If the decimal number is greater than 1, move the decimal point to the left by the appropriate number of places, and add that number from the exponent of 10.

Example	Exercise 1
Simplify $\left(4.2 \times 10^8\right)\left(1.8 \times 10^{12}\right)$. Express your answer in scientific notation.	Simplify $\left(2.0 \times 10^{-18}\right)\left(3.4 \times 10^5\right)$. Express your answer in scientific notation.
SOLUTION: First multiply 4.2 by 1.8. $$4.2 \cdot 1.8 = 7.56$$ Use the product rule of exponents to multiply the powers of 10. $$10^8 \cdot 10^{12} = 10^{8+12} = 10^{20}$$ So, $$\left(4.2 \times 10^8\right)\left(1.8 \times 10^{12}\right) = 7.56 \times 10^{20}$$ The result is in scientific notation.	

Exercise 2

Simplify $\left(1.598\times10^{-14}\right)\div\left(4.7\times10^{9}\right)$. Express your answer in scientific notation.

Exercise 3

Simplify $\left(3.286\times10^{16}\right)\div\left(6.2\times10^{-6}\right)$. Express your answer in scientific notation.

Exercise 4

Simplify $\left(6.02\times10^{23}\right)\left(5.5\times10^{-4}\right)$. Express your answer in scientific notation.

Exercise 5

Simplify $\left(6.0\times10^{-15}\right)\div\left(7.5\times10^{-10}\right)$. Express your answer in scientific notation.

4.2.3 Solve applications involving scientific notation.

To solve applications involving scientific notation, begin by determining which operation to perform.

Many problems involving scientific notation will require you to multiply or divide two numbers. Multiply if one quantity is being counted repeatedly. Divide if a quantity is being broken into equally sized quantities.

Follow the procedure for multiplying or dividing numbers in scientific notation.

For example, if the average annual sales at a chain restaurant is $\$2.5 \times 10^6$ per restaurant, what are the annual sales if there are 2000 locations of that restaurant?

To solve this problem, multiply the sales per restaurant by the number of restaurants. (2000 restaurants can be written in scientific notation as 2.0×10^3. Multiply the two decimal numbers, and add the exponents for the powers of 10.

$$\left(2.5 \times 10^6\right) \cdot \left(2.0 \times 10^3\right) = 5.0 \times 10^9$$

If a problem calls for you to add or subtract two decimal numbers, consider changing the two numbers to standard notation before adding or subtracting.

Example	Exercise 1
If a computer can perform a calculation in 8.0×10^{-10} second, how long will it take to perform 4.8×10^{13} calculations?	If a computer can perform a calculation in 8.0×10^{-10} second, how many calculations can it perform in 90 seconds?
SOLUTION: Multiply the time required for one calculation by the number of calculations. $$\left(8.0 \times 10^{-10}\right) \cdot \left(4.8 \times 10^{13}\right) = 38.4 \times 10^3$$ $$= 3.84 \times 10^4$$ It would take 3.84×10^4 or 38,400 seconds.	

Exercise 2

The speed of light is 1.86×10^5 miles per second. If Pluto is 4,555,000,000 miles from the sun, how long does it take light from the sun to reach Pluto? Round to the nearest second.

Exercise 3

The speed of light is 1.86×10^5 miles per second. How far can light from the sun travel in 30 minutes?

Exercise 4

A social media site averages 3.0×10^8 posts per day. How many posts would appear on the site in 1 year?

Exercise 5

The mass of a hydrogen atom is 1.66×10^{-24} grams. What is the mass of 25,000,000,000,000,000 atoms of hydrogen?

4.3.1   Identify parts of a polynomial (coefficient, term, degree, factor, constant).

A **polynomial** in a single variable x is a sum of **terms** of the form ax^n, where a is a real number and n is a whole number.

The **coefficient** of a term is the numerical factor of the term. When listing the coefficient of a term be sure to include its sign.

The **degree** of each term in a polynomial in a single variable is equal to the variable's exponent.

The **degree of a polynomial** in a single variable is equal to the greatest degree of any of its terms.

A polynomial is written in **descending order** if its terms are listed from greatest degree to least degree.

A term that does not have a variable factor is called a **constant term**. The degree of a constant term is 0.

Example	Exercise 1
For the polynomial $x^2 - 49$, identify each term, the coefficient of each term, and the degree of each term.	For the polynomial $7x^6 - 4x^3 - x$, identify each term, the coefficient of each term, and the degree of each term.
SOLUTION: There are two terms: x^2 and -49. The coefficient of the x^2 term is 1, because x^2 is equivalent to $1x^2$. The coefficient of the second term is -49. The degree of the x^2 term is 2, because its exponent is 2 The degree of the second term is 0 because it is a constant term.	

Exercise 2

For the polynomial $45x^8$, identify each term, the coefficient of each term, and the degree of each term.

Exercise 3

Rewrite the polynomial $-4x^2 + 9x^3 + 7 - 16x$ in descending order. Identify the leading coefficient and the degree of the polynomial.

Exercise 4

Rewrite the polynomial $25 - 9x - x^4$ in descending order. Identify the leading coefficient and the degree of the polynomial.

Exercise 5

Rewrite the polynomial $5x^6 - x^2 + 3x^4 + 21$ in descending order. Identify the leading coefficient and the degree of the polynomial.

4.3.2 Classify polynomials.

A polynomial can be classified by the number of its terms. A term in a polynomial is either a constant term or the product of a constant and variables.

A polynomial with one term is called a **monomial**.

Examples of monomials: $\qquad 7x \qquad\qquad 9x^2 \qquad\qquad -18x^7$

A polynomial with two terms is called a **binomial**.

Examples of binomials: $\qquad 3x-8 \qquad\qquad x^2-100 \qquad\qquad 8x^5-21x^3$

A polynomial with three terms is called a **trinomial**.

Examples of trinomials: $\qquad x^2+5x+6 \qquad\qquad 5x^2-6x-25 \qquad\qquad x^3+x-40$

We do not use special names for polynomials with four or more terms.

Example	**Exercise 1**
Identify the polynomial as a monomial, binomial, or trinomial: $5x^2-125$	Identify the polynomial as a monomial, binomial, or trinomial: $x^2-19x+90$
SOLUTION: There are two terms: $5x^2$ and -125. A polynomial with two terms is called a binomial.	

Exercise 2
Identify the polynomial as a monomial, binomial, or trinomial: $10x^5$

Exercise 3
Identify the polynomial as a monomial, binomial, or trinomial: $3x^4 - 18x^2 - 30$

Exercise 4
Identify the polynomial as a monomial, binomial, or trinomial: $2x^9$

Exercise 5
Identify the polynomial as a monomial, binomial, or trinomial: $x^2 - 24x$

4.3.3 Evaluate polynomial expressions.

To evaluate a polynomial for a particular value of a variable, substitute the value for the variable in the polynomial and simplify the resulting expression by following the order of operations.

When substituting into the polynomial, replace each occurrence of the variable with a set of parentheses before substituting the value for the variables. This helps to avoid sign mistakes when working with exponents and negative numbers.

For example, suppose you needed to evaluate x^2 for $x = -3$. Begin by rewriting x^2 as $(\ )^2$, then write -3 inside the parentheses.

$$(-3)^2 = 9$$

If you did not use parentheses, you might write down -3^2 and that would produce an incorrect result of -9.

The expression x^2 is equivalent to $x \cdot x$. When x is to be replaced by -3, this produces $(-3)(-3)$ or 9.

In general, using parentheses is a great way to organize your work while avoiding sign mistakes.

Example	Exercise 1
Evaluate $x^2 + 7x + 21$ for $x = -6$. SOLUTION: Replace each occurrence of the variable x with a set of parentheses. Substitute -6 in the parentheses for each variable. Simplify the resulting expression by following the order of operations. $$x^2 + 7x + 21$$ $$(-6)^2 + 7(-6) + 21$$ $$= 36 + 7(-6) + 21$$ $$= 36 - 42 + 21$$ $$= 15$$	Evaluate $x^2 - 5x - 19$ for $x = 8$.

Exercise 2

Evaluate $3x^2 - 4x - 25$ for $x = 7$.

Exercise 3

Evaluate $x^3 - 4x^2 + 9x + 35$ for $x = -2$.

Exercise 4

Evaluate $x^4 - 7x^3 - 5x^2$ for $x = 10$.

Exercise 5

Evaluate $-x^6 - 64$ for $x = -4$.

4.3.4   Add polynomials.

Two polynomials can be added by combining like terms.

Recall that two terms are like terms if they have the same variable part – the same variables with the same exponents.

To combine like terms, add their coefficients and keep the variable part.

If a term in one polynomial does not have a like term in the other polynomial, simply copy it down in the sum.

$$\left(x^3 - 3x^2 - x + 9\right) + \left(8x^2 - 9x - 20\right)$$
$$= x^3 - 3x^2 - x + 9 + 8x^2 - 9x - 20$$
$$= x^3 + \left(-3x^2 + 8x^2\right) + \left(-x - 9x\right) + \left(9 - 20\right)$$
$$= x^3 + 5x^2 - 10x - 11$$

Some students find it convenient when adding polynomials to write like terms vertically.

Example	Exercise 1
Add $\left(2x^2 + 3x - 8\right) + \left(4x^2 - 4x - 19\right)$.	Add $\left(x^2 + 2x + 5\right) + \left(x^2 + 4x + 15\right)$.
SOLUTION: The expression can be rewritten without parentheses. Then combine like terms. $\left(2x^2 + 3x - 8\right) + \left(4x^2 - 4x - 19\right)$ $= 2x^2 + 3x - 8 + 4x^2 - 4x - 19$ Combine $2x^2$ and $4x^2$ to produce $6x^2$. Combine $3x$ and $-4x$ to produce $-x$. Combine -8 and -19 to produce -27. $\left(2x^2 + 3x - 8\right) + \left(4x^2 - 4x - 19\right)$ $= 2x^2 + 3x - 8 + 4x^2 - 4x - 19$ $= 6x^2 - x - 27$	

Exercise 2

Add $\left(x^2 + 7x - 18\right) + \left(x^2 - 3x - 21\right)$.

Exercise 3

Add $\left(x^2 + 6x + 5\right) + \left(x^2 + x - 16\right)$.

Exercise 4

Add $\left(x^2 + 3x + 4\right) + \left(x^2 - 3x + 21\right)$.

Exercise 5

Add $\left(2x^2 + 7x - 10\right) + \left(3x^2 - 2x + 19\right)$.

4.3.5   Subtract polynomials.

The procedure for subtracting polynomials is similar to the procedure for adding polynomials, with one difference.

Begin by distributing a factor of -1 to each term in the polynomial that is being subtracted. This will basically change the sign of each term in the polynomial that is being subtracted.

After distributing -1 to the polynomial being subtracted, proceed by combining like terms.

$$\left(x^3 - 3x^2 - x + 9\right) - \left(8x^2 - 9x - 20\right)$$
$$= x^3 - 3x^2 - x + 9 - 8x^2 + 9x + 20$$
$$= x^3 - 11x^2 + 8x + 29$$

Example	**Exercise 1**
Subtract $\left(x^2 + 8x - 17\right) - \left(3x^2 + 5x + 20\right)$.	Subtract $\left(5x^2 - 2x - 19\right) - \left(2x^2 - 11x + 42\right)$.
SOLUTION: The expression can be rewritten without parentheses by distributing -1 to each term in the polynomial that is being subtracted. Then combine like terms. $\left(x^2 + 8x - 17\right) - \left(3x^2 + 5x + 20\right)$ $= x^2 + 8x - 17 - 3x^2 - 5x - 20$ $= -2x^2 + 3x - 37$	

Exercise 2

Subtract $\left(x^2 - 3x + 2\right) - \left(x^2 - 8x + 40\right)$.

Exercise 3

Subtract $\left(x^3 + 2x^2 - 3x - 4\right) - \left(7x^2 - 10x + 26\right)$.

Exercise 4

Subtract $\left(4x^2 + 4x - 30\right) - \left(x^2 - 4x + 30\right)$.

Exercise 5

Subtract $\left(x^4 + 7x + 52\right) - \left(6x^3 - x^2 - 19x\right)$.

4.3.6   Multiply monomials.

Being able to multiply monomials is essential in order to be able to multiply polynomials.

Begin by multiplying the coefficients. This will be the coefficient of the product.

Next, multiply variable factors by using the product rule for exponents.

Product Rule for Exponents
For any base x, $x^m \cdot x^n = x^{m+n}$.

For example, consider $8x^8 y^5 z^2 \cdot 3x^{15} y$.
- Coefficients: $8 \cdot 3 = 24$
- x: $x^8 \cdot x^{15} = x^{8+15} = x^{23}$
- y: $y^5 \cdot y = y^{5+1} = y^6$
- z: z^2 remains unchanged since z is only a factor of one monomial.

So, $8x^8 y^5 z^2 \cdot 3x^{15} y = 24 x^{23} y^6 z^2$.

Example	Exercise 1
Simplify $5x^7 \cdot 2x^{11}$.	Simplify $2x^8 \cdot x^7$.
SOLUTION: Multiply the coefficients: $$5 \cdot 2 = 10$$ Multiply the variable factors by using the product rule for exponents: $$x^7 \cdot x^{11} = x^{7+11} = x^{18}$$ So, $5x^7 \cdot 2x^{11} = 10x^{18}$.	

Exercise 2

Simplify $-9x \cdot 3x^6$.

Exercise 3

Simplify $4x^{10} \cdot 3x^{20} \cdot 2x^{30}$.

Exercise 4

Simplify $7x^2y \cdot 8x^8y^{17}$.

Exercise 5

Simplify $-6x^3y^2z^4 \cdot 2x^6z^{11}$.

4.3.7 Multiply a monomial and a polynomial.

To multiply a monomial and a polynomial, apply the distributive property.

Distributive Property
For any real numbers a, b, and c,

$$a(b+c) = ab + ac$$

After distributing the monomial to each term in the polynomial, simplify each product by using the same approach used to multiply monomials.

- Multiply the coefficients.
- Multiply variable factors by applying the product rule for exponents.
  For any base x, $x^m \cdot x^n = x^{m+n}$.

For example, consider $7x^2(x^2 - 3x + 10)$.

Distribute $7x^2$ to each term of the polynomial: $7x^2 \cdot x^2 - 7x^2 \cdot 3x + 7x^2 \cdot 10$
Simplify each term: $7x^4 - 21x^3 + 70x^2$

Example	**Exercise 1**
Multiply $2x^3(4x^2 - 5x - 6)$.	Multiply $x^5(x^4 + 3x^2 + 11x)$.
SOLUTION: Distribute the monomial $2x^3$ to each term of the polynomial $4x^2 - 5x - 6$. Then simplify each product by using the same approach that is used to multiply two monomials. $2x^3(4x^2 - 5x - 6)$ $= 2x^3 \cdot 4x^2 - 2x^3 \cdot 5x - 2x^3 \cdot 6$ $= 8x^5 - 10x^4 - 12x^3$	

Exercise 2

Multiply $8x\left(3x^2 + 2x - 20\right)$.

Exercise 3

Multiply $-10x^4\left(7x^2 - 2x\right)$.

Exercise 4

Multiply

$-3x^3\left(2x^5 - 3x^4 - 4x^3 - 8x^2 - 21x - 100\right)$.

Exercise 5

Multiply $9x^3 y^2\left(x^7 y^5 - 4x^4 y - 11xy^{14}\right)$.

4.3.8   Multiply polynomials.

To multiply two polynomials, distribute each term in the first polynomial to each term in the second polynomial.

For two binomials of the form $a+b$ and $c+d$,
$$(a+b)(c+d) = a\cdot c + a\cdot d + b\cdot c + b\cdot d$$

After distributing each term in the first polynomial to each term in the second polynomial, simplify each product by using the same approach used to multiply monomials.

- Multiply the coefficients.
- Multiply variable factors by applying the product rule for exponents.
  For any base x, $x^m \cdot x^n = x^{m+n}$.

Finish by combining any like terms.

Example	Exercise 1
Multiply $(5x+2)(6x+7)$.	Multiply $(x+2)(x+15)$.
SOLUTION: Distribute each term in the first polynomial, $5x+2$, to each term in the second polynomial. Simplify each product, and finish by combining any like terms. $(5x+2)(6x+7)$ $= 5x\cdot 6x + 5x\cdot 7 + 2\cdot 6x + 2\cdot 7$ $= 30x^2 + 35x + 12x + 14$ $= 30x^2 + 47x + 14$	

Exercise 2

Multiply $(x+16)(x-11)$.

Exercise 3

Multiply $(3x+10)(2x-17)$.

Exercise 4

Multiply $(7x+6)(2x-13)$.

Exercise 5

Multiply $(x+6)(x^2-7x-20)$.

4.3.9 Multiply the sum and difference of two terms.

Suppose you were to multiply the sum of two terms $(a+b)$ by the difference of the same two terms $(a-b)$. Using the procedure for multiplying polynomials,

$$(a+b)(a-b)$$
$$=a^2-ab+ab-b^2$$
$$=a^2-b^2$$

The formula for multiplying the sum and difference of two terms is given by
$$(a+b)(a-b)=a^2-b^2$$

Two binomials of the form $a+b$ and $a-b$ are also called conjugates.

If you forget the formula for multiplying the sum and difference of two terms, you can always multiply them as you would multiply any two polynomials.

Example	Exercise 1
Multiply $(3x+2)(3x-2)$.	Multiply $(x+6)(x-6)$.
SOLUTION: Use the formula $(a+b)(a-b)=a^2-b^2$. In this case, $a=3x+2$ and $b=3x-2$. $$(3x+2)(3x-2)=(3x)^2-(2)^2$$ $$=9x^2-4$$ The same result can be found without using the formula by multiplying the two polynomials. $$(3x+2)(3x-2)$$ $$=3x\cdot3x-3x\cdot2+2\cdot3x-2\cdot2$$ $$=9x^2-6x+6x-4$$ $$=9x^2-4$$	

Exercise 2

Multiply $(x-19)(x+19)$.

Exercise 3

Multiply $(8x+11)(8x-11)$.

Exercise 4

Multiply $(x^4+1)(x^4-1)$.

Exercise 5

Multiply $(5x^2+7)(5x^2-7)$.

4.3.10 Square binomials.

A binomial $a + b$ can be squared by multiplying it by itself.
Using the procedure for multiplying polynomials,

$$\begin{aligned}(a+b)^2 &= (a+b)(a+b) \\ &= a^2 + ab + ab + b^2 \\ &= a^2 + 2ab + b^2\end{aligned}$$

The formula for squaring a binomial $a + b$ is given by

$$(a+b)^2 = a^2 + 2ab + b^2$$

In a similar fashion, the formula for squaring a binomial $a - b$ is given by

$$(a-b)^2 = a^2 - 2ab + b^2$$

In each case, identify a and b, substitute into the formula, and simplify the expression.

If you forget the formula for squaring a binomial, you can always multiply the binomial by itself.

Example	Exercise 1
Simplify $(2x + 9)^2$.	Simplify $(x + 4)^2$.
SOLUTION: Use the formula $(a+b)^2 = a^2 + 2ab + b^2$. In this case, $a = 2x$ and $b = 9$. $$\begin{aligned}(2x+9)^2 &= (2x)^2 + 2(2x)(9) + (9)^2 \\ &= 4x^2 + 36x + 81\end{aligned}$$ The same result can be found without using the formula by multiplying $2x + 9$ by itself. $$\begin{aligned}(2x+9)&(2x+9) \\ &= 2x \cdot 2x + 2x \cdot 9 + 9 \cdot 2x + 9 \cdot 9 \\ &= 4x^2 + 18x + 18x + 81 \\ &= 4x^2 + 36x + 81\end{aligned}$$	

Exercise 2 Simplify $(x-9)^2$.	**Exercise 3** Simplify $(5x-6)^2$.
Exercise 4 Simplify $(7x+13)^2$.	**Exercise 5** Simplify $(11x-8)^2$.

4.4.1 Factor out the GCF of a polynomial.

The first step for factoring any polynomial is to factor out the GCF (or greatest common factor) of its terms.

Begin by finding the GCF.
- The GCF of the coefficients is the largest integer that divides into each coefficient evenly.
- For a variable factor to be included in the GCF, it must be a factor of each term of the polynomial.
  If a variable is a factor of each term, that variable's exponent in the GCF will be equal to the smallest exponent that it has in any of the terms. (This is what the terms share in common.)

Write the GCF in front of a set of parentheses.

To determine the polynomial that will be written inside the parentheses, determine what polynomial the GCF must be multiplied by in order to equal the original polynomial.
This polynomial can also be found by dividing each term in the original polynomial by the GCF.

You can check your work by multiplying the GCF by the polynomial inside the parentheses.

Example	Exercise 1
Factor out the GCF: $6x^{10} - 15x^9 + 21x^7$.	Factor out the GCF: $18x^9 - 66x^5$.
SOLUTION: The GCF of 6, 15, and 21 is 3. The GCF of x^{10}, x^9, and x^7 is x^7. Factoring out the GCF begins as $6x^{10} - 15x^9 + 21x^7 = 3x^7 (\quad - \quad + \quad)$ To find the three terms inside the parentheses, determine what to multiply $3x^7$ by in order to equal the original polynomial. (You can also divide each of the original polynomial's terms by $3x^7$.) $6x^{10} - 15x^9 + 21x^7 = 3x^7 \left(2x^3 - 5x^2 + 7 \right)$	

Exercise 2

Factor out the GCF: $8x^5 + 12x^3 - 28$.

Exercise 3

Factor out the GCF: $27x^7 - 36x^6 - 81x^5$.

Exercise 4

Factor out the GCF: $20x^7 - 50x^6 + 10x^3$.

Exercise 5

Factor out the GCF: $6x^7y^3 - 14x^5y^5 - 20x^3y^2$.

4.4.2 Factor trinomials with a leading coefficient of 1.

A trinomial of degree 2 with a leading coefficient is a trinomial that can be written in the form $x^2 + bx + c$, where b and c are integers.
Examples of trinomials of degree 2 with a leading coefficient of 1:

$$x^2 + 12x + 32 \qquad x^2 - 9x + 14 \qquad x^2 + 8x - 20 \qquad x^2 - x - 12$$

If $x^2 + bx + c$, where b and c are integers, is factorable, it can be factored as the product of two binomials of the form $(x + m)(x + n)$, where m and n are integers.

To be able to factor $x^2 + bx + c$ into the product $(x + m)(x + n)$, you must find the correct integers m and n.
Since $(x + m)(x + n)$ has a product of $x^2 + (m + n)x + mn$, we observe that $m + n$ must be equal to b and mn must be equal to c.

In other words, m and n are two integers whose product is c and whose sum is b.

If the values of m and n do not come to you quickly, try listing all of the factor pairs of c first. This will make it easier to find the pair of integers whose sum is b.

Keep in mind that some polynomials cannot be factored and are called prime. If two integers m and n do not exist such that $mn = c$ and $m + n = b$, the trinomial $x^2 + bx + c$ is prime.

Example	Exercise 1
Factor completely: $x^2 + 12x - 28$	Factor completely: $x^2 - 15x + 54$
SOLUTION:	
Since the constant term is negative, m and n must have opposite signs.	
Find two numbers whose product is -28 and whose sum is 12.	
Factor pairs of -28:	
$-1 \cdot 28,\ 1(-28),\ -2 \cdot 14,\ 2(-14),\ -4 \cdot 7,\ 4(-7)$	
The integers -2 and 14 have a product of -28 and a sum of 12, so $m = -2$ and $n = 14$.	
$$x^2 + 12x - 28 = (x - 2)(x + 14)$$	
The order of the factors can be reversed.	
$$x^2 + 12x - 28 = (x + 14)(x - 2)$$	

Exercise 2

Factor completely: $x^2 + 13x - 30$

Exercise 3

Factor completely: $x^2 - 13x + 30$

Exercise 4

Factor completely: $x^2 + 20x + 91$

Exercise 5

Factor completely: $x^2 - 18x - 280$

4.4.3 Factor trinomials with a leading coefficient other than 1.

Examples of trinomials of degree 2 with a leading coefficient other than 1:

$$3x^2 + 13x + 12 \qquad 5x^2 - 41x + 8 \qquad 10x^2 - 23x - 42 \qquad 3x^2 - 10x - 48$$

If $ax^2 + bx + c$, where a, b, and c are integers, is factorable, it can be factored as the product of two binomials of the form $(\text{Variable Term} + \text{Constant})(\text{Variable Term} + \text{Constant})$.

The two constant terms must have a product of ax^2, and the two constant terms must have a product of c.

Once you find all the possible variable terms and constants, use trial and error to find the factored form that produces a middle term of bx when the two binomials are multiplied.

Consider the trinomial $3x^2 + 13x + 12$. The only factor pair of $3x^2$ is $x \cdot 3x$. The factor pairs of 12 are $1 \cdot 12$, $2 \cdot 6$, and $3 \cdot 4$.

Possibilities for the factored form of $3x^2 + 13x + 12$:

$$(x+1)(3x+12) \qquad\qquad (x+2)(3x+6) \qquad\qquad (x+3)(3x+4)$$
$$(x+12)(3x+1) \qquad\qquad (x+6)(3x+2) \qquad\qquad (x+4)(3x+3)$$

The correct factored form of $3x^2 + 13x + 12$ is $(x+3)(3x+4)$, which can be verified by multiplying the two binomials.

$$(x+3)(3x+4) = 3x^2 + 4x + 9x + 12 = 3x^2 + 13x + 12$$

There is another approach for factoring this type of trinomial that uses factoring by grouping. There are videos for this alternate approach in your MyLab Math course.

Example	Exercise 1
Factor completely: $2x^2 - x - 21$	Factor completely: $3x^2 + 17x + 10$
SOLUTION: The only factor pair of $2x^2$ is $x \cdot 2x$. The factor pairs of -21 are $-1 \cdot 21$, $1(-21)$, $-3 \cdot 7$, and $3(-7)$. Use trial and error to find the pairing that produces a middle term of $-x$. The factored form is $\qquad 2x^2 - x - 21 = (x+3)(2x-7)$ This result can be checked by multiplying the two binomials. $\qquad (x+3)(2x-7) = 2x^2 - 7x + 6x - 21$ $\qquad\qquad\qquad\quad = 2x^2 - x - 21$	

Exercise 2
Factor completely: $5x^2 - 39x + 54$

Exercise 3
Factor completely: $6x^2 - 5x - 6$

Exercise 4
Factor completely: $6x^2 + 7x - 24$

Exercise 5
Factor completely: $8x^2 + 24x - 80$

4.4.4   Factor polynomials by grouping.

Factoring by grouping is an effective strategy when working with polynomials that have four or more terms.

Factoring a Polynomial with Four Terms by Grouping
- Factor a common factor out of the group of the first two terms.
- Factor a common factor out of the group of the last two terms.
- If the two groups share a common binomial factor, this binomial can be factored out of the two groups to complete the factoring of the polynomial.

Consider the polynomial $3x^3 + 27x^2 + 5x + 45$. The GCF of the first two terms is $3x^2$, and the GCF of the last two terms is 5. Factor the GCF from each group of two terms.
$$3x^3 + 27x^2 + 5x + 45 = 3x^2(x+9) + 5(x+9)$$

The two groups have a common binomial factor of $(x+9)$. This common binomial factor can now be factored out of each of the two groups.
$$3x^3 + 27x^2 + 5x + 45 = 3x^2(x+9) + 5(x+9)$$
$$= (x+9)(3x^2 + 5)$$

If the two groups do not have a common binomial factor, consider reordering the terms and trying again.
Occasionally a negative common factor will need to be factored out of a group in order for the binomial factors in each group to be exactly the same.

Example	**Exercise 1**
Factor by grouping: $x^3 + 8x^2 + 7x + 56$.	Factor by grouping: $5x^3 - 20x^2 + 9x - 36$.
SOLUTION: The first two terms have a GCF of x^2. Factor that from the first two terms. $x^3 + 8x^2 + 7x + 56 = x^2(x+8) + 7x + 56$ The last two terms have a GCF of 7. Factor 7 from the last two terms. $x^3 + 8x^2 + 7x + 56 = x^2(x+8) + 7(x+8)$ Factor out the common binomial factor of $(x+8)$. $x^3 + 8x^2 + 7x + 56 = x^2(x+8) + 7(x+8)$ $= (x+8)(x^2 + 7)$	

Exercise 2

Factor by grouping: $x^3 - 15x^2 - 13x + 195$.

Exercise 3

Factor by grouping: $4x^3 - 56x^2 + x - 14$.

Exercise 4

Factor by grouping: $4x^3 - 20x^2 + 11x - 55$.

Exercise 5

Factor by grouping: $6x^2 - 21x + 10x - 35$.

4.4.5   Factor a difference of squares.

A difference of squares is a binomial that can be expressed in the form $a^2 - b^2$.

A difference of squares can be factored using the formula $a^2 - b^2 = (a+b)(a-b)$.

The formula can be verified by finding the product of the binomials $(a+b)(a-b)$.
$$(a+b)(a-b) = a^2 - ab + ab - b^2$$
$$= a^2 - b^2$$

To identify a binomial as a difference of squares you must verify that each term is a perfect square.
- Variable factors must have exponents that are multiples of 2, such as x^2, y^4, and z^6.
- Any constant must be a perfect square.
  Here are the first ten perfect squares:
  $1^2 = 1$, $2^2 = 4$, $3^2 = 9$, $4^2 = 16$, $5^2 = 25$, $6^2 = 36$, $7^2 = 49$, $8^2 = 64$, $9^2 = 81$, $10^2 = 100$

Before attempting to factor a binomial as a difference of squares, determine whether a GCF can be factored out of the two terms.

A sum of squares cannot be factored in general.

When factoring a difference of squares, the binomial factor that is a difference can be factored further.

Example	**Exercise 1**
Factor $x^2 - 169$.	Factor $x^2 - 225$.
SOLUTION: Begin by determining if each term can be written as a perfect square. The first term is a perfect square: $x^2 = (x)^2$. The second term is a perfect square, since $169 = (13)^2$. So, the binomial can be rewritten as a difference of squares. $$x^2 - 169 = (x)^2 - (13)^2$$ Use the formula $a^2 - b^2 = (a+b)(a-b)$ to complete the factoring. $$x^2 - 169 = (x)^2 - (13)^2$$ $$= (x+13)(x-13)$$	

Exercise 2

Factor $9x^2 - 100$.

Exercise 3

Factor $36x^6 - 49y^4$.

Exercise 4

Factor $10x^2 - 1440$.

Exercise 5

Factor $x^4 - 81$.

4.4.6   Factor the sum or difference of two cubes.

A sum of cubes is a binomial that can be expressed in the form $a^3 + b^3$.

A sum of cubes can be factored using the formula $a^3 + b^3 = (a+b)(a^2 - ab + b^2)$.

The formula can be verified by finding the product of the polynomials $(a+b)(a^2 - ab + b^2)$.

$$(a+b)(a^2 - ab + b^2) = a^3 - a^2b + ab^2 + a^2b - ab^2 + b^3$$
$$= a^3 + b^3$$

To identify a binomial as a sum of cubes you must verify that each term is a perfect cube.

- Variable factors must have exponents that are multiples of 3, such as x^3, y^6, and z^9.
- Any constant must be a perfect cube.
  Here are the first ten perfect cubes: 1, 8, 27, 64, 125, 216, 343, 512, 729, 1000

A difference of cubes is a binomial that can be expressed in the form $a^3 - b^3$, and can be factored using the formula $a^3 - b^3 = (a-b)(a^2 + ab + b^2)$.

Notice that the two formulas are very similar, with the only difference involving signs.

Sum of Cubes: $a^3 + b^3 = (a+b)(a^2 - ab + b^2)$

Difference of Cubes: $a^3 - b^3 = (a-b)(a^2 + ab + b^2)$

Before attempting to factor a binomial as a difference of cubes, determine whether a GCF can be factored out of the two terms.

Example	Exercise 1
Factor $x^3 + 343$.	Factor $x^3 - 216$.
SOLUTION: Begin by determining if each term can be written as a perfect cube. The first term is a perfect cube: $x^3 = (x)^3$. The second term is a perfect cube, since $343 = 7^3$. So, the binomial can be rewritten as a sum of cubes. $$x^3 + 343 = (x)^3 + (7)^3$$ Use the formula $a^3 + b^3 = (a+b)(a^2 - ab + b^2)$ to complete the factoring. $$x^3 + 343 = (x)^3 + (7)^3$$ $$= (x+7)(x^2 - x \cdot 7 + 7^2)$$ $$= (x+7)(x^2 - 7x + 49)$$	

Exercise 2

Factor $27x^3 - 125$.

Exercise 3

Factor $4x^3 + 500$.

Exercise 4

Factor $x^3 - 512y^3$.

Exercise 5

Factor $343x^3 + 1000y^3$.

4.4.7 Factor polynomials completely.

Factoring Polynomials
1. Factor out any common factors.
2. Determine the number of terms in the polynomial.
    a. If there are only two terms, check to see if the binomial is a difference of squares, a sum of cubes, or a difference of cubes.
        i. Difference of Squares: $a^2 - b^2 = (a+b)(a-b)$
        ii. Sum of Cubes: $a^3 + b^3 = (a+b)(a^2 - ab + b^2)$
        iii. Difference of Cubes: $a^3 - b^3 = (a-b)(a^2 + ab + b^2)$
    b. If there are three terms, determine if the trinomial has degree 2. If so, the strategy will depend on whether the leading coefficient is 1 or some number other than 1.
        i. $x^2 + bx + c = (x+m)(x+n)$

        Find two integers m and n with a product of c and a sum of b.
        ii. $ax^2 + bx + c$, where $a \neq 1$

        List the factor pairs of ax^2 and c, then use trial and error to find the two binomials that produce a middle term of bx.
    c. If there are four or more terms, try factoring by grouping.
3. After the polynomial has been factored, make sure that any factor with two or more terms does not have any common factors that can be factored out. Also check to see if any factor can be factored further.
4. Check your factoring through multiplication.

Example	Exercise 1
Factor completely $2x^2 - 29x + 90$.	Factor completely $8x^4 - 12x^3 - 2x$.
SOLUTION: There is no common factor that can be factored out of the three terms, so this trinomial has a leading coefficient other than 1.	
Factor pairs of $2x^2$: $x \cdot 2x$ Factor pairs of 90: $1 \cdot 90$, $2 \cdot 45$, $3 \cdot 30$, $5 \cdot 18$, $6 \cdot 15$, $9 \cdot 10$	
Both constant terms must be negative to produce a middle term of $-29x$. Use trial and error to find the correct factored form. $$2x^2 - 29x + 90 = (x-10)(2x-9)$$	

Exercise 2

Factor completely $x^2 - 13x - 48$.

Exercise 3

Factor completely $25x^2 - 121$.

Exercise 4

Factor completely $x^3 - 4x^2 + 11x - 44$.

Exercise 5

Factor completely $x^3 + 729$.

4.5.1   Solve quadratic equations by factoring.

A **quadratic equation** is an equation that can be written as $ax^2 + bx + c = 0$, where a, b, and c are real numbers and $a \neq 0$. This form is the **standard form of a quadratic equation**.

Solving a quadratic equation by factoring uses the zero-factor property of real numbers.

Zero-Factor Property of Real Numbers
If $a \cdot b = 0$, then $a = 0$ or $b = 0$.

Solving Quadratic Equations by Factoring
1. Write the equation in standard form: $ax^2 + bx + c = 0$.
    Collect all terms on one side of the equation.
    It helps to collect all of the terms so that the coefficient of the squared term is positive.

2. Factor the expression $ax^2 + bx + c$ completely.

3. Apply the zero-factor property of real numbers and set each factor equal to 0.
    Solve the resulting equations.
    Each of these equations should be a linear equation.

Example	Exercise 1
Solve $x^2 + 5x - 84 = 0$.	Solve $x^2 - 16x + 28 = 0$.
SOLUTION: Since the trinomial is set equal to 0, factor the trinomial and set each factor equal to 0. $$x^2 + 5x - 84 = 0$$ $$(x+12)(x-7) = 0$$ $$x+12 = 0 \quad \text{or} \quad x-7 = 0$$ $$x = -12 \quad \text{or} \quad x = 7$$ $$\{-12, 7\}$$	

Exercise 2

Solve $x^2 - 49 = 0$.

Exercise 3

Solve $x^2 + 6x = 2x + 252$.

Exercise 4

Solve $x(x-2) = 10x + 45$.

Exercise 5

Solve $2x^2 - 23x + 30 = 0$.

4.5.2   Solve quadratic equations by the square root property.

The equation $x^2 = 49$ has two solutions, $x = -7$ and $x = 7$.

Although this equation can be solved by factoring, the equation can also be solved by taking the square root of both sides of the equation. In order to obtain both solutions, it is necessary to take both the positive and negative square roots of 49. This is written as $\pm\sqrt{49}$.

$$x^2 = 49$$
$$\sqrt{x^2} = \pm\sqrt{49}$$
$$x = \pm 7$$
$$\{-7, 7\}$$

This approach can be used to solve any equation that contains only a squared term and constant terms.

Solving Quadratic Equations by the Square Root Property
1. Isolate the squared term.

2. Take the square root of each side of the equation.
    Be sure to take both the positive and negative $(\pm)$ square root of the constant.

3. Simplify each square root.

4. Solve by isolating the variable.

Example	Exercise 1
Solve $(x+2)^2 - 7 = 17$.	Solve $(x-8)^2 = -25$.
SOLUTION: Isolate the squared term, then take the square root of each side of the equation. $$(x+2)^2 - 7 = 17$$ $$(x+2)^2 = 24$$ $$\sqrt{(x+2)^2} = \pm\sqrt{24}$$ $$x + 2 = \pm 2\sqrt{6}$$ $$x = -2 \pm 2\sqrt{6}$$ $$\{-2 - 2\sqrt{6}, -2 + 2\sqrt{6}\}$$	

Exercise 2

Solve $(x+3)^2 - 7 = 53$.

Exercise 3

Solve $(2x+1)^2 - 11 = 25$.

Exercise 4

Solve $(3x-7)^2 + 18 = -30$.

Exercise 5

Solve $2(x+3)^2 - 9 = 27$

4.5.3   Solve quadratic equations by completing the square.

The goal of solving a quadratic equation by completing the square is to take the quadratic equation and rewriting it in a form that can be solved by using the square root property. In order to do that you will need a perfect square trinomial.

A binomial $x^2 + bx$ can be rewritten as a perfect square trinomial by adding the constant $\left(\dfrac{b}{2}\right)^2$ to it.

Consider the binomial $x^2 - 18x$. Half of -18 is -9, and when squared that equals 81.
Add 81 to $x^2 - 18x$ and factor.
$$x^2 - 18x + 81 = (x-9)(x-9) = (x-9)^2$$

Completing the Square
1. Isolate all variable terms on one side of the equation, with the constant term on the other side of the equation.
2. Identify the coefficient of the first-degree term. Take half of that number, square it, and add that to both sides of the equation.
3. Factor the resulting perfect square trinomial.
4. Take the square root of each side of the equation. Be sure to include $\pm$ on the side where the constant is.
5. Solve the resulting equation.

In order to complete the square, the coefficient of the squared term must be 1.
If the equation is of the form $ax^2 + bx + c = 0$, begin by dividing both sides of the equation by a.
Then solve by completing the square.

Example	Exercise 1
Solve $x^2 + 6x - 23 = 0$ by completing the square. SOLUTION: Begin by isolating the variable terms. Then complete the square by adding $\left(\dfrac{6}{2}\right)^2$ or 9 to both sides of the equation. Finish by using the square root property. $$x^2 + 6x - 23 = 0$$ $$x^2 + 6x = 23$$ $$x^2 + 6x + 9 = 23 + 9$$ $$(x+3)^2 = 32$$ $$\sqrt{(x+3)^2} = \pm\sqrt{32}$$ $$x + 3 = \pm 4\sqrt{2}$$ $$x = -3 \pm 4\sqrt{2}$$ $$\left\{-3 - 4\sqrt{2}, -3 + 4\sqrt{2}\right\}$$	Solve $x^2 - 4x - 96 = 0$ by completing the square.

Exercise 2

Solve $x^2 + 8x + 24 = 0$ by completing the square.

Exercise 3

Solve $x^2 - 2x + 10 = 0$ by completing the square.

Exercise 4

Solve $x^2 - 5x - 3 = 0$ by completing the square.

Exercise 5

Solve $2x^2 + 6x - 3 = 0$ by completing the square.

4.5.4   Solve quadratic equations by using the quadratic formula.

Another strategy for solving equation is by using the **quadratic formula**.

This is a formula that can be used to solve any quadratic equation, once it is written in standard form $\left(ax^2 + bx + c = 0\right)$.

For a general quadratic equation $ax^2 + bx + c = 0$ (where a, b, and c are real numbers and $a \neq 0$), the quadratic formula tells us that the solutions of the equation are given by

$$x = \frac{-b \pm \sqrt{b^2 - 4ac}}{2a}$$

To solve a quadratic equation using the quadratic formula:
- Write the equation in standard form, $ax^2 + bx + c = 0$.
- Identify the coefficients a, b, and c.
- Substitute into the quadratic formula.
- Simplify the radicand and the denominator.
- Simplify the square root.
- Simplify the fraction.

Example	Exercise 1
Solve $x^2 - 8x + 20 = 0$ by using the quadratic formula.	Solve $x^2 + 36x + 320 = 0$ by using the quadratic formula.
SOLUTION: The equation is in standard form, with $a = 1$, $b = -8$, and $c = 20$. Substitute into the quadratic formula. $x = \dfrac{-b \pm \sqrt{b^2 - 4ac}}{2a}$ $x = \dfrac{-(-8) \pm \sqrt{(-8)^2 - 4(1)(20)}}{2(1)}$ $x = \dfrac{8 \pm \sqrt{-16}}{2}$ $x = \dfrac{8 \pm 4i}{2}$ $x = 4 \pm 2i$	

Exercise 2

Solve $x^2 + 6x - 9 = 0$ by using the quadratic formula.

Exercise 3

Solve $x^2 + 2x - 80 = 0$ by using the quadratic formula.

Exercise 4

Solve $2x^2 - 5x - 20 = 0$ by using the quadratic formula.

Exercise 5

Solve $3x^2 + 8x + 2 = 0$ by using the quadratic formula.

4.5.5   Distinguish between linear and quadratic models in real world situations.

Some quantities can be modeled by a linear function, and others can be modeled by a quadratic function.

If you can determine that the quantity is related to a constant rate of change, like when each item is associated with a particular value, then the function can be modeled by a linear function. An example of this would be a scenario when you are trying to model the weekly pay for a worker who works x hours and is paid $20/hour. The function would be $f(x) = 20x$.

The height of a projectile after x seconds can be modeled by a quadratic function.

If you have a set of paired data, plot the points (x, y).
- If the data follow a linear pattern, the quantity y can be modeled by a linear function of x.
- If the data follow a U-shaped pattern (like the shape of a parabola), or some part of a U-shaped pattern, the quantity y can be modeled by a quadratic function of x.

Example	**Exercise 1**
A publisher charges $1.50 for each copy of its newspaper that it sells. Would the revenue function for selling x copies of the newspaper be best modeled by a linear function or a quadratic function?	A projectile is launched straight upward from the ground with an initial velocity of 60 feet/second. Would the function for the height of the projectile after x seconds be best modeled by a linear function or a quadratic function?
SOLUTION: Since the revenue increases by $1.50 for every copy sold, the revenue can be modeled by the linear function $f(x) = 1.50x$.	

Exercise 2

The distance that a skydiver falls (y), in feet, is recorded after x seconds.

Time (seconds)	0.25	0.5	1	1.3	1.75	2
Distance (ft)	1	4	16	27	49	64

Would the function for the distance after x seconds be best modeled by a linear function or a quadratic function?

Exercise 3

The distance required for a car to stop when traveling at different speeds is listed in the table.

Speed (mph)	20	30	40	50	60	70
Stopping Distance (ft)	20	45	79	125	181	246

Would the function for the stopping distance for a car traveling at x mph be best modeled by a linear function or a quadratic function?

Exercise 4

The Fahrenheit temperature (y) and the number of cricket chirps (x) recorded in 15 seconds are measured.

Chirps	22	17	30	52	29	40	58	33
Temp. (°F)	65	55	70	92	70	79	98	75

Would the function for the temperature in degrees Fahrenheit when x chirps are measured in 15 seconds be best modeled by a linear function or a quadratic function?

Exercise 5

A company has made a new smart phone and needs to determine the optimal price to charge for the phone. Its research department has predicted total profit (y, in $ billions) for various potential prices (x).

Price	500	600	700	800	900	1000
Profit ($ billions)	1	1.3	1.5	1.6	1.6	1.5

Would the function for total profit if the price is set at x dollars be best modeled by a linear function or a quadratic function?

4.5.6   Solve applications that involve quadratic equations and models (including the Pythagorean theorem).

Problems involving the product of consecutive integers or the area of a rectangle often produce equations that are quadratic.

Two consecutive integers can be represented by x and $x+1$.
Two consecutive odd integers can be represented by x and $x+2$.
Two consecutive even integers can also be represented by x and $x+2$.

The area of a rectangle is given by the formula Area $=$ Length $\cdot$ Width or $A = LW$.

The **Pythagorean theorem** relates the three sides of a right triangle. A **right triangle** is a triangle that has a right angle. The measure of a right angle is $90°$.
The side opposite the right angle is called the **hypotenuse**. The other two sides are called **legs**.

Pythagorean Theorem
For any triangle whose hypotenuse has length c and whose legs have lengths a and b, respectively,
$$a^2 + b^2 = c^2$$

For applications involving the Pythagorean theorem, start by drawing the triangle described in the problem.
Label each side, then substitute into the Pythagorean theorem and solve the resulting equation.

Example	Exercise 1
Two consecutive positive integers have a product of 210. Find the integers.	Two consecutive positive even integers have a product of 80. Find the integers.

Example (continued)

SOLUTION:
Let x represent the first integer and $x+1$ represent the second integer.
The product of the two integers is $x(x+1)$.
Set the product equal to 210 and solve.

$$x(x+1) = 210$$
$$x^2 + x = 210$$
$$x^2 + x - 210 = 0$$
$$(x+15)(x-14) = 0$$
$$x+15 = 0 \quad \text{or} \quad x-14 = 0$$
$$x = -15 \quad \text{or} \quad x = 14$$

Since the integers are positive, omit $x = -15$.
The first integer (x) is 14.
The second integer $(x+1)$ is 15.

Exercise 2
The width of a rectangle is 3 inches less than its length. If the area of the rectangle is 154 square inches, find the length and width.

Exercise 3
A rectangular garden covers 136 square meters. If the length of the garden is 1 meter more than twice its width, find the length and the width.

Exercise 4
A 10-foot ladder is resting against a wall that is 8 feet above the ground. How far is the bottom of the ladder from the base of the wall?

Exercise 5
A 50-foot wire connects the top of a utility pole to a point on the ground. If the height of the pole is 7 feet more than the distance from the base of the pole to the point where the wire is anchored in the ground, how tall is the pole? (Round to the nearest tenth of a foot.)

5.1.1 Calculate sales tax, total price, and sale price.

Sales tax is computed as a percentage of the sale price.
The tax rate is determined by the local city, county, or state.

The sales tax can be computed by multiplying the tax rate by the sale price.

$$\text{Sales Tax} = \text{Tax Rate} \cdot \text{Sale Price}$$

The sale price can also be found from this equation by dividing the sales tax by the tax rate.

If the tax rate is known, it must be expressed as a decimal when used in the equation.

For example, suppose that a state charges 8% sales tax. If a gas grill has a sales price of $699, the sales tax can be computed as follows:

$$\text{Sales Tax} = 0.08(\$699) = \$55.92$$

The total price of an item can be found by adding the sales tax to the sale price.
For the gas grill mentioned above, the total price is equal to $\$699 + \$55.92$, or $754.92.

Example	Exercise 1
A state has an 8% sales tax. Find the sales tax on a bicycle whose price is $485.	If a state has a 7% sales tax, find the sales tax on a computer that sells for $599.
SOLUTION: To compute the sales tax, find 8% of 485 by multiplying its decimal equivalent (0.08) by 485. $$0.08(\$485) = \$38.80$$ The sales tax is $38.80.	

Exercise 2
A state has an 8.5% sales tax. If the sales tax on a pair of athletic shoes is $7.48, what was the original price of the shoes?

Exercise 3
If a state has a 9.25% sales tax and the sales tax charged on a saw is $4.07, find the original price of the saw.

Exercise 4
A store is selling a sweater for $78. If the state adds on a 6.5% sales tax, find the total price of the sweater.

Exercise 5
A furniture store is selling a sofa for $800. If the state adds on a 7.5% sales tax, find the total price of the sofa.

5.1.2 Calculate commission.

Some salespeople are paid by commission.
Commission on a sale is calculated as a percentage of the sale price.
Commission rates vary by industry and company.
Commission is typically paid by the seller out of the sale price.

The amount of commission can be computed by multiplying the commission rate by the sale price.

$$\text{Commission} = \text{Commission Rate} \cdot \text{Sale Price}$$

The sale price can also be found from this equation by dividing the commission by the commission rate.

If the commission rate is known, it must be expressed as a decimal when used in the equation.

For example, suppose a salesperson earns 2% commission on each sale. If the salesperson sells a condo for $270,000, the commission can be computed as follows:

$$\text{Commission} = 0.02(\$270,000) = \$5400$$

Example	Exercise 1
A realtor earns a 3% commission on all sales. If her total sales last year amounted to $3,455,900, find the amount she earned in commission. SOLUTION: The amount earned in commission is 3% of $3,455,900. $$0.03(\$3,455,900) = \$103,677$$ She earned $103,677 in commission.	A realtor earns a 6% commission on all sales if the realtor represents both the buyer and seller of a house. Find the commission earned by a realtor if the selling price was $349,000 and the realtor was the agent of both the buyer and seller.

Exercise 2

A ticket agent adds a 10% commission for concert tickets. If the original price of a concert ticket was $125, how much would the agent earn as a commission?

Exercise 3

A representative at a local gym earns a 5% commission for each $250 annual membership sold. After paying out the commission, how much does the gym keep from the sale of an annual membership?

Exercise 4

A homeowner paid a 6% commission to the realtors for selling her home. If the sale price was $275,000, how much did the homeowner have left after paying the commission?

Exercise 5

A car dealer marks up its cars by 5%. The dealer pays a salesman a 25% commission on the net profit for each car sold. If the dealer paid $27,000 for a car, how much profit did it make after marking the car up by 5% and paying a 25% commission on the net profit?

5.1.3 Calculate tips.

A tip is an amount of money added to a bill in recognition of good service.
The tip is computed as a percentage of the bill.
The tip rate often depends on the quality of the service.

The tip amount can be computed by multiplying the tip rate by the amount of the bill.

$$\text{Tip Amount} = \text{Tip Rate} \cdot \text{Total Bill}$$

The amount of the bill can also be found from this equation by dividing the tip by the tip rate.

The tip rate must be expressed as a decimal when used in the equation.

For example, suppose that a diner decides to add an 18% tip to the bill for outstanding service. If the bill was $75, the tip can be computed as follows:

$$\text{Tip} = 0.18(\$75) = \$13.50$$

The total amount paid can be found by adding the tip to the bill.
For the diner mentioned above, the total amount paid is equal to $75 + \$13.50$, or $88.50.

Example	**Exercise 1**
A decent tip to a waiter or waitress for good service at a restaurant is 15% of the bill. If the total bill for a dinner is $89.20, compute a 15% tip. SOLUTION: To compute the tip, find 15% of $89.20. $$0.15(\$89.20) = \$13.38$$ The tip is $13.38.	A diner pays a 20% tip when the waiter or waitress provides outstanding service. If the service was outstanding, compute a 20% tip on a total bill of $37.65.

Exercise 2
If a diner left a $7.50 tip for a $60 dinner, what percent of the bill was the tip?

Exercise 3
A diner leaves an 18% tip on a bill of $64.50. What is the total bill after the tip is added?

Exercise 4
If a customer leaves a 25% tip on a bill of $92.68, find the total bill after the tip is added.

Exercise 5
A pizza delivery restaurant adds a $5 delivery charge to each order. If a customer orders $35 worth of pizzas and adds a 10% tip to the total bill after the delivery charge is added, what is the total amount spent by the customer?

5.1.4 Calculate original price, discount, total cost, tax, and markup.

A store can adjust their prices in one of three ways: applying a discount, adding tax, and adding a markup.

Discount
Stores lower the price on items in the form of a discount.
The discount is typically computed as a percentage of the sale price of the item.

$$\text{Discount} = \text{Discount Rate} \cdot \text{Sale Price}$$

Tax
Stores add sales tax to the price of an item.
The tax is typically computed as a percentage of the sale price of the item.

$$\text{Tax} = \text{Tax Rate} \cdot \text{Sale Price}$$

Markup
Stores compute the price they will sell an item by adding a markup to the price they paid for the item.
The markup is typically computed as a percentage of the price the store paid.

$$\text{Markup Amount} = \text{Markup Rate} \cdot \text{Cost}$$

Example	**Exercise 1**
A bookstore is having a 20% off sale. Find the discount on a book whose original price was $27.50. SOLUTION: The discount is 20% of $27.50. $$0.2(\$27.50) = \$5.50$$ The discount is $5.50.	A restaurant is offering a 10% discount on all desserts. If a dessert is priced at $6.50, find the discount on the dessert.

Exercise 2

A clothing store has marked down all of its prices by 15%. If the original price for a suit was $450, find the sale price.

Exercise 3

A clothing store is having a sale on jeans and has marked down all jeans by 40%. If the original price for a pair of jeans is $96, find the sale price.

Exercise 4

An appliance store is having a 25% off sale. After the discount is taken, a 9% sales tax is added on to the sale price. Find the total price, including sales tax, of a refrigerator whose original price was $1400.

Exercise 5

A hardware store is having a 20% off sale on all tools. In addition, members of their club save an extra 10% after the discount is applied. If the original price of a drill is $87, find the sale price after the 20% discount and then the 10% club member discount is applied.

5.2.1 Compute simple interest.

Interest is an amount that a borrower pays for the privilege of using someone else's money.

Simple interest is computed once, and covers the entire length of the loan.

The equation for computing simple interest, I, is $I = P \cdot r \cdot t$.
- P is the principal, which is the amount being borrowed.
- r is the interest rate.
  The interest rate is given as a percent, but must be converted to a decimal for use in the formula.
- t is the time, in years.
  If the time is less than 1 year, express it as a fraction of a year.

For example, suppose $500 is borrowed at a simple annual interest rate of 8% for 2 years.
The principal P is $500. The interest rate r is 8%, or 0.08. The time t is 2 years.

$$I = P \cdot r \cdot t$$
$$I = \$500 \cdot 0.08 \cdot 2$$
$$I = \$80$$

Example	Exercise 1
Bruce borrows $500 at 6% annual simple interest. How much interest does he pay if he pays the loan off after one year? SOLUTION: To compute the simple interest, use the formula $I = P \cdot r \cdot t$, with $P = \$500$, $r = 0.06$, and $t = 1$. $$I = P \cdot r \cdot t$$ $$I = (\$500)(0.06)(1)$$ $$I = \$30$$ The interest after one year is $30.	A borrower takes out a loan of $4000, agreeing to simple annual interest of 12%. How much interest is accrued after one year?

Exercise 2

Tina takes out a loan of $280, agreeing to pay a simple annual interest rate of 8%. If she pays the loan back in 9 months, how much interest does she owe?

Exercise 3

Jackie borrows $2500 at a simple annual interest rate of 3%. If she pays back the loan after $3\frac{1}{2}$ years, how much interest does she have to pay on the loan?

Exercise 4

Hank borrows $3000 at a simple annual interest rate of 5%. If he pays the loan back in 2 months, how much interest does he owe?

Exercise 5

A borrower takes out an $800 loan at a simple annual interest rate of 10%. If it takes the borrower 9 years to pay back the loan, how much interest is owed?

5.2.2 Find principal, interest rate, or time in the simple interest formula.

Interest is an amount that a borrower pays for the privilege of using someone else's money.

Simple interest is computed once, and covers the entire length of the loan.

The equation for simple interest is $I = P \cdot r \cdot t$.
- I is the amount of simple interest.
- P is the principal, which is the amount being borrowed.
- r is the interest rate.
  The interest rate is given as a percent, but must be converted to a decimal for use in the formula.
- t is the time, in years.
  If the time is less than 1 year, express it as a fraction of a year.

If three of the four quantities are known, you can use the equation to solve for the fourth quantity.

Build a table with the four variables: I, P, r, and t. Fill in the table with values extracted from the problem. Use a variable for the unknown quantity.

If the unknown is the interest rate, after you solve for the rate convert it from a decimal to a percent.

Example	Exercise 1
Steve took out a loan with a simple annual interest rate of 6%. If he repaid the loan in 1 year and built up an interest charge of $72, what was the amount that Steve borrowed?	A borrower took out a loan at 12% simple annual interest, and repaid the loan after 9 months. If the total interest paid was $252, what was the principal of the loan?
SOLUTION: Use the formula $I = P \cdot r \cdot t$. Substitute $72 for I, 0.06 for r, and 1 for t. Solve the resulting equation for the principal, P. $$I = P \cdot r \cdot t$$ $$72 = P(0.06)(1)$$ $$72 = 0.06P$$ $$\frac{72}{0.06} = \frac{0.06P}{0.06}$$ $$1200 = P$$ Steve borrowed $1200.	

Exercise 2
Priscilla borrowed $1000 at a simple annual interest rate. If she repaid the loan after 2 years and paid a total of $125 in interest, what was the interest rate of her loan?

Exercise 3
A borrower took out a loan of $700 at a simple annual interest rate. The loan was repaid after 6 months, and the total interest was $28. What was the interest rate of the loan?

Exercise 4
Tonya borrowed $200 at 9% simple annual interest. If she paid a total of $54 in interest, how long did it take her to repay the loan?

Exercise 5
A borrower took out a loan of $750 at 2% annual interest. If she paid a total of $10 in interest, how many months did it take her to pay back the loan?

5.2.3 Solve applications involving simple interest.

Interest is an amount that a borrower pays for the privilege of using someone else's money.

Simple interest is computed once, and covers the entire length of the loan.

The equation for simple interest is $I = P \cdot r \cdot t$.
- I is the amount of simple interest.
- P is the principal, which is the amount being borrowed.
- r is the interest rate.
  The interest rate is given as a percent, but must be converted to a decimal for use in the formula.
- t is the time, in years.
  If the time is less than 1 year, express it as a fraction of a year.

If three of the four quantities are known, you can use the equation to solve for the fourth quantity.

Build a table with the four variables: I, P, r, and t. Fill in the table with values extracted from the problem. Use a variable for the unknown quantity.

If the unknown is the interest rate, after you solve for the rate convert it from a decimal to a percent.

The total amount that is paid back is equal to the principal plus the interest.

Example	Exercise 1
Sara borrowed $750 at a simple annual interest rate of 5%. If she repaid the loan after 1 year, find the total amount repaid including the interest.	A borrower took out a loan of $1000 at a simple annual rate of 6%. If the loan was repaid after 9 months, find the total amount repaid including the interest.
SOLUTION: First find the interest using the formula $I = P \cdot r \cdot t$. Substitute $750 for P, 0.05 for r, and 1 for t. Solve the resulting equation for the amount of interest, I. $$I = P \cdot r \cdot t$$ $$I = 750(0.05)(1)$$ $$I = \$37.50$$ Sara would have to pay back the principal ($750) plus the interest ($37.50). $$\$750 + \$37.50 = \$787.50$$ Sara has to repay $787.50.	

Exercise 2

Tyrese borrowed $10,000 for 1 year at a simple annual interest rate of 5%. He also loaned out $8000 for 18 months at a simple annual interest rate of 8%. Find Tyrese's net profit from these two loans.

Exercise 3

A borrower took out a loan of $5000 at a simple annual interest rate of 2%. If the borrower pays the loan back in 12 equal monthly payments, find the amount of each monthly payment.

Exercise 4

Isabela borrowed $100 at a simple annual interest rate of 4% for 6 months. If she pays it back in 6 equal monthly payments, find the amount of each monthly payment.

Exercise 5

A borrower took out a loan of $800 at a simple annual interest rate of 10% for two years. If she decides to pay the loan back in 24 equal monthly payments, find the amount of each monthly payment.

5.3.1   Compute compound interest.

Many savings accounts pay interest using **compound interest**.
Compound interest is interest that is computed at regular intervals.
Interest on the principal is computed during the first period. During the second period, interest is paid on both the principal and the interest earned during the first period.

If P dollars are deposited in an account that pays an annual interest rate r that is compounded n times per year, the amount A in the account after t years is given by the formula

$$A = P\left(1 + \frac{r}{n}\right)^{nt}$$

The amount P is referred to as the principal.

The annual interest rate r is often reported as a percent, but it must be converted to a decimal for this formula.

The amount of interest earned can be found by subtracting the principal P from the balance A.

Common Values of n
Monthly: $n = 12$ Quarterly: $n = 4$ Semiannually: $n = 2$ Annually: $n = 1$

Example	Exercise 1
If \$2500 is deposited in an account that pays 2% annual interest, compounded quarterly, find the amount of interest earned in 6 years. SOLUTION: First compute the balance in 6 years using the formula $A = P\left(1 + \frac{r}{n}\right)^{nt}$ with $P = \$2500$, $r = 0.02$, $n = 4$ times per year, and $t = 6$ years. $A = P\left(1 + \frac{r}{n}\right)^{nt}$ $A = 2500\left(1 + \frac{0.02}{4}\right)^{4 \cdot 6}$ $A = 2500(1.005)^{24}$ $A \approx 2817.90$ The interest is found by subtracting the principal from the balance after 6 years: $\$2817.90 - \$2500 = \$317.90$ The interest earned is \$317.90.	If \$1000 is deposited in an account that pays 5% annual interest, compounded annually, find the balance after 4 years.

Exercise 2
If $800 is deposited in an account that pays 8% annual interest, compounded semiannually, find the balance after 10 years.

Exercise 3
If $15,000 is deposited in an account that pays 6% annual interest, compounded monthly, find the balance after 12 years.

Exercise 4
If $50,000 is deposited in an account that pays 3% annual interest, compounded monthly, find the amount of interest earned in 5 years.

Exercise 5
If $4750 is deposited in an account that pays 1.3% annual interest, compounded semiannually, find the amount of interest earned in 6 years.

5.3.2   Find principal, interest rate, or time in the compound interest formula.

Compound Interest Formula

If P dollars are deposited in an account that pays an annual interest rate r that is compounded n times per year, the amount A in the account after t years is given by the formula

$$A = P\left(1+\frac{r}{n}\right)^{nt}$$

If the principal P is unknown, you can solve for P by using $P = \dfrac{A}{\left(1+\dfrac{r}{n}\right)^{nt}}$.

If the time t is unknown, you can solve for t by using $t = \dfrac{\ln\left(\dfrac{A}{P}\right)}{n\cdot\ln\left(1+\dfrac{r}{n}\right)}$.

If the interest rate r is unknown, you can solve for r by using $r = n\left(\left(\dfrac{A}{P}\right)^{\frac{1}{nt}}-1\right)$.

Example	Exercise 1
$100 was deposited in an account that pays 9% annual interest, compounded monthly. How long will it take for the balance to reach $250? Round to the nearest tenth of a year. SOLUTION: Substitute in the formula for compound interest. Use logarithms to solve for the time t. $$A = P\left(1+\frac{r}{n}\right)^{nt}$$ $$250 = 100\left(1+\frac{0.09}{12}\right)^{12t}$$ $$250 = 100(1.0075)^{12t}$$ $$2.5 = 1.0075^{12t}$$ $$\ln 2.5 = \ln 1.0075^{12t}$$ $$\ln 2.5 = 12t\cdot \ln 1.0075$$ $$\frac{\ln 2.5}{12\ln 1.0075} = t$$ $$t \approx 10.2$$ It would take 10.2 years.	$20,000 was deposited in an account that pays 4% annual interest, compounded quarterly. How long will it take to earn $5000 in interest? Round to the nearest tenth of a year.

Exercise 2 Find the principal needed to generate a balance of $4000 in 5 years, if it is deposited in an account that pays 6% annual interest that is compounded monthly. Round to the nearest dollar.	**Exercise 3** Find the principal needed to generate a balance of $1000 in 2 years, if it is deposited in an account that pays 15% annual interest that is compounded quarterly. Round to the nearest dollar.
Exercise 4 $5000 was deposited in an account that paid interest that was compounded semiannually, and in 10 years the balance had grown to $10,000. Find the annual interest rate, rounded to the nearest tenth of a percent.	**Exercise 5** $200 was deposited in an account that paid interest that was compounded quarterly, and in 4 years the balance had grown to $300. Find the annual interest rate, rounded to the nearest tenth of a percent.

5.3.3 Solve applications involving compound interest.

If P dollars are deposited in an account that pays an annual interest rate r that is compounded n times per year, the amount A in the account after t years is given by the formula

$$A = P\left(1 + \frac{r}{n}\right)^{nt}$$

To solve applied problems, substitute values for four of the variables in the formula and solve for the unknown variable.

If the amount A is unknown, substitute into the formula and simplify the resulting expression on the right side of the formula.

If the principal P is unknown, you can solve for P by using $P = \dfrac{A}{\left(1 + \dfrac{r}{n}\right)^{nt}}$.

If the time t is unknown, you can solve for t by using $t = \dfrac{\ln\left(\dfrac{A}{P}\right)}{n \cdot \ln\left(1 + \dfrac{r}{n}\right)}$.

If the interest rate r is unknown, you can solve for r by using $r = n\left(\left(\dfrac{A}{P}\right)^{\frac{1}{nt}} - 1\right)$.

Example	Exercise 1
Adriana inherited $10,000, which she was not allowed to access for 5 years. If the inheritance was put in a savings account that paid 2% annual interest compounded quarterly, how much will Adriana have in 5 years? SOLUTION: Substitute 10,000 for P, 0.02 for r, 4 for n, and 5 for t in the formula $A = P\left(1 + \dfrac{r}{n}\right)^{nt}$. Solve for A, rounding to two decimal places. $A = P\left(1 + \dfrac{r}{n}\right)^{nt}$ $A = 10,000\left(1 + \dfrac{0.02}{4}\right)^{4 \cdot 5}$ $A = 10,000(1.005)^{20}$ $A = 11,048.96$ Adriana will have $11,048.96 in 5 years.	Tzu-Wei wants to generate $2000. How much does he need to invest in a fund that pays 7% annual interest, compounded annually, in order to earn $2000 in interest in the next 4 years?

Exercise 2 An investor deposits $5000 in an account that pays 9% annual interest, compounded quarterly. How long will it take for the balance to reach $10,000?	**Exercise 3** A bank offers two types of accounts. Account A pays 2% annual interest, compounded quarterly. Account B pays 2.2% annual interest, compounded annually. If $100 is invested in each account for 6 years, which account will have the greater balance?
Exercise 4 Deshanay wants to help a friend buy a car. Deshanay borrows $3500 for 3 years from her bank at 4% annual interest compounded semiannually, and then lends the $3500 to her friend. Deshanay loans her friend the $3500 for 3 years and charges her friend 6% annual interest, compounded annually. How much net profit will Deshanay have after the loans are repaid?	**Exercise 5** An investor deposits $10,000 in an account that pays interest that is compounded quarterly. What annual interest rate is required if the investor wants to have a balance of $12,500 after 4 years? Round to the nearest tenth of a percent.

6.1.1 Identify relations and functions.

A **relation** is a rule that takes an input value from one set and assigns a particular output value from another set to it.

A relation for which each input value is assigned one and only one output value is called a **function**.

The **domain** of the function is the set of input values, and the **range** is the set of output values.

To determine if a relation is a function, you must verify that each input in the domain is only associated with one output in the domain.

If you can find an input value that is associated with more than one output value, then you have shown that the relation is not a function.

For example, suppose a relation that took a student at your college as an input and listed the year the student first enrolled at your college. This relation is a function because each student has one particular year that they first enrolled.

However, if a relation is defined in the reverse order, then this relation is not a function because one particular year has many students that first enrolled in that year.

Example	Exercise 1
Would a relation that took a U.S. governor's name as an input and listed that governor's state as an output be a function? Why or why not? SOLUTION: This would be a function, because each governor is the governor of only one state.	Would a relation that took a letter grade as an input and listed a student from your school that earned that grade on their last exam as an output be a function? Why or why not?

Exercise 2

Would a relation that took a month as its input and listed a person whose birthday is in that month as an output be a function? Why or why not?

Exercise 3

Would a relation that took a student in your class as an input and listed the year they were born as an output be a function? Why or why not?

Exercise 4

Is the relation defined by the set of ordered pairs $\{(-2,2),(-1,1),(0,0),(1,1),(2,2)\}$ a function? Why or why not?

Exercise 5

Is the relation defined by the set of ordered pairs $\{(4,-2),(1,-1),(0,0),(1,1),(4,2)\}$ a function? Why or why not?

6.1.2   Evaluate functions using function notation.

Function notation is a way to present the output value of a function for the input x.

Suppose that a function took an input, multiplied it by 3, and then added 10 to that. This can be expressed in function notation as $f(x) = 3x + 10$.

Analyzing $f(x) = 3x + 10$

- The variable x inside the parentheses is the input variable for the function.
- The name of the function is f.
- $f(x)$ represents the output value of the function f when the input value is x.
- The expression on the right side, $3x + 10$, is the formula for the output of this function.

To **evaluate** a function for a particular value of the variable, substitute the value for the variable in the function's formula and then simplify the resulting expression.

Suppose that for the function $f(x) = 3x + 10$ you wanted to evaluate the function for $x = -8$. In other words, you wanted to find $f(-8)$. Substitute -8 for x and simplify.

$$f(x) = 3x + 10$$
$$f(-8) = 3(-8) + 10 \qquad \text{Substitute } -8 \text{ for } x.$$
$$= -24 + 10 \qquad \text{Multiply.}$$
$$= -14 \qquad \text{Simplify.}$$

Example	Exercise 1
For $f(x) = 6x + 33$, find $f(5)$.	For $f(x) = -8x - 11$, find $f\left(-\dfrac{3}{4}\right)$.
SOLUTION: Substitute 5 for x and simplify. $$f(x) = 6x + 33$$ $$f(5) = 6(5) + 33$$ $$= 30 + 33$$ $$= 63$$	

Exercise 2

For $f(x) = x^2 + 6x - 20$, find $f(-4)$.

Exercise 3

For $f(x) = x^2 - 7x - 22$, find $f(-5)$.

Exercise 4

For $f(x) = x^2 + 20x - 45$, find $f(13)$.

Exercise 5

For $f(x) = \dfrac{x+6}{3x-2}$, find $f(2)$.

6.1.3 Find the domain and range of a function.

The domain of a function is its set of possible input values. The domain of a function can be read from left to right along the *x*-axis.

The range of a function is its set of possible output values. The range of a function can be read vertically from the bottom to the top of the graph along the *y*-axis.

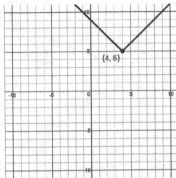

Created using the Desmos Graphing Calculator

The graph extends indefinitely to the left and to the right, so its domain is the set of all real numbers $(-\infty, \infty)$.

The function has a minimum value of 5, so the range is $[5, \infty)$.

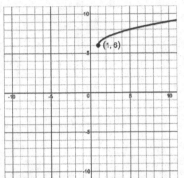

Created using the Desmos Graphing Calculator

The graph begins at the point $(1, 6)$ and extends indefinitely to the right, so its domain $[1, \infty)$.

The function has a minimum value of 6, so the range is $[6, \infty)$.

Example	**Exercise 1**
Find the domain and range of the function that has been graphed.	Find the domain and range of the function that has been graphed.

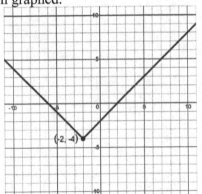

Created using the Desmos Graphing Calculator

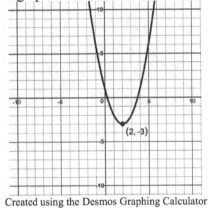

Created using the Desmos Graphing Calculator

SOLUTION:
The graph extends indefinitely to the left and to the right, so its domain is the set of all real numbers $(-\infty, \infty)$.

The function has a minimum value of -4, so the range is $[-4, \infty)$.

Exercise 2

Find the domain and range of the function that has been graphed.

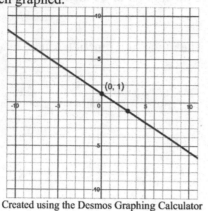

Created using the Desmos Graphing Calculator

Exercise 3

Find the domain and range of the function that has been graphed.

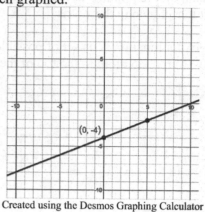

Created using the Desmos Graphing Calculator

Exercise 4

Find the domain and range of the function that has been graphed.

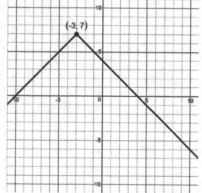

Created using the Desmos Graphing Calculator

Exercise 5

Find the domain and range of the function that has been graphed.

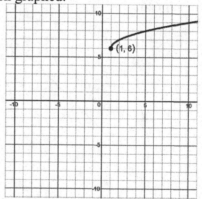

Created using the Desmos Graphing Calculator

6.1.4 Use the vertical line test to determine if a graph is a function.

The Vertical Line Test
If a vertical line can be drawn that intersects a graph at more than one point, the graph does not represent a function.

If a vertical line (of the form $x = a$) intersected a graph at more than one point, that would indicate that one input value of x is associated with two or more output values.
This violates the definition of a function (each input can only be associated with one output).

To apply the vertical line test, look for a vertical line that would intersect the graph more than once.

For example, for the graph below, the vertical line $x = 6$ intersects the graph at two points: $(6, -2)$ and $(6, 5)$. Therefore, the graph does not represent a function.

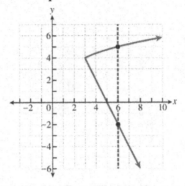

Example	**Exercise 1**
Use the vertical-line test to determine whether the graph represents a function.	Use the vertical-line test to determine whether the graph represents a function.
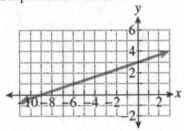	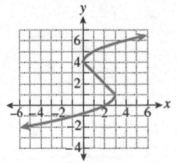
SOLUTION: Since a vertical line cannot be drawn that intersects the graph at more than one point, the graph does represent a function.	

Exercise 2
Use the vertical-line test to determine whether the graph represents a function.

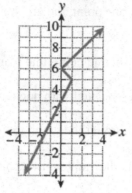

Exercise 3
Use the vertical-line test to determine whether the graph represents a function.

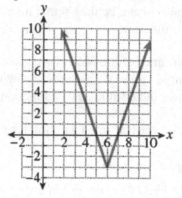

Exercise 4
Use the vertical-line test to determine whether the graph represents a function.

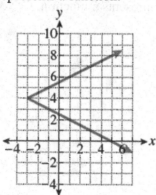

Exercise 5
Use the vertical-line test to determine whether the graph represents a function.

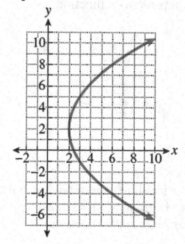

6.2.1 Identify linear functions.

A **linear function** is a function that can be written in the form $f(x) = mx + b$.

A linear function in the variable x will not have the variable x in a denominator, and the variable x cannot have an exponent other than 1.

- $f(x) = 3x^2 - 7$ Not a linear function; the variable has an exponent other than 1.

- $f(x) = \dfrac{3}{x} - 7$ Not a linear function; the variable is in a denominator.

- $f(x) = 3x - 7$ Linear function; it has the form $f(x) = mx + b$ with $m = 3$ and $b = -7$.

- $f(x) = -7$ Linear function; it has the form $f(x) = mx + b$ with $m = 0$ and $b = -7$.

Example	Exercise 1
Is the function $f(x) = \dfrac{3}{4}x - 8$ a linear function? SOLUTION: The variable x is not in a denominator, and it does not have an exponent other than 1. This function is a linear function. It is of the form $f(x) = mx + b$, with $m = \dfrac{3}{4}$ and $b = -8$.	Is the function $f(x) = x^3 - 8$ a linear function?

Exercise 2

Is the function $f(x) = \dfrac{3x-8}{x+4}$ a linear function?

Exercise 3

Is the function $f(x) = 8x$ a linear function?

Exercise 4

Is the function $f(x) = 8$ a linear function?

Exercise 5

Is the function $f(x) = 7 - 2x$ a linear function?

6.2.2 Evaluate linear functions.

Evaluating a linear function is similar to evaluating any function.

To evaluate a function for a particular value of the variable, substitute the value for the variable in the function's formula and then simplify the resulting expression.

Suppose that for the function $f(x) = -6x + 5$ you wanted to find $f(15)$.
Substitute 15 for x and simplify.

$$f(x) = -6x + 5$$
$$f(15) = -6(15) + 5$$
$$= -90 + 5$$
$$= -85$$

Example	Exercise 1
A minor league baseball team hosts events for companies. The cost, in dollars, for hosting an event with x attendees is given by the function $f(x) = 22x + 100$. Use the function to find the cost for an event with 75 attendees. SOLUTION: To find the cost for an event with 75 attendees, substitute 75 for x in the function and simplify. $$f(x) = 22x + 100$$ $$f(75) = 22(75) + 100$$ $$= 1650 + 100$$ $$= 1750$$ The cost for an event with 75 attendees is $1750.	An estimate of the outdoor Fahrenheit temperature can be estimated from the number of times (x) a cricket chirps in a 15-second interval by using the function $f(x) = 1.1x + 40.6$. If a cricket chirps 25 times in 15 seconds, use the function to estimate the Fahrenheit temperature.

Exercise 2

The Fahrenheit temperature associated with a Celsius temperature x is given by the function $f(x) = \frac{9}{5}x + 32$. Use the function to find the Fahrenheit temperature associated with a Celsius temperature of 100°C.

Exercise 3

The Celsius temperature associated with a Fahrenheit temperature x is given by the function $f(x) = \frac{5(x-32)}{9}$. Use the function to find the Celsius temperature associated with a Fahrenheit temperature of 50°F.

Exercise 4

The number y of students pursuing a graduate degree in mathematical sciences x years after 2010 can be described by the function $f(x) = 700x + 23,100$. Use the function to predict the number of students that will be pursuing a graduate degree in mathematical sciences in 2025.

Exercise 5

A student has started selling smartphone cases to support her education. The profit from selling x cases can be described by the function $f(x) = 4x - 500$. Use the function to find the profit earned if she sells 350 cases.

6.2.3   Interpret the graph of a linear function (domain, range, slope, intercepts).

A linear function can be written in the form $f(x) = mx + b$, and its graph is a non-vertical line.

The domain of a linear function is the set of all real numbers, $(-\infty, \infty)$, as a linear function's graph extends indefinitely to the left and to the right.

The range of a linear function is also the set of all real numbers, $(-\infty, \infty)$, unless the function is a constant function whose graph is a horizontal line. The range of a constant function of the form $f(x) = b$ is $\{b\}$.

To find the slope of a linear function from its graph, first find the coordinates of two points (x_1, y_1) and (x_2, y_2) on the line. Then use the slope formula, $m = \dfrac{y_2 - y_1}{x_2 - x_1}$, to compute the slope.

The x-intercept is the point where the graph of the linear function intersects the x-axis. The x-intercept has a y-coordinate of 0 and will be of the form $(a, 0)$.

The y-intercept is the point where the graph of the linear function intersects the y-axis. The y-intercept has an x-coordinate of 0 and will be of the form $(0, b)$.

Example	**Exercise 1**
Find the slope and any intercepts for the linear function from its graph.	Find the slope and any intercepts for the linear function from its graph.
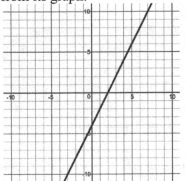	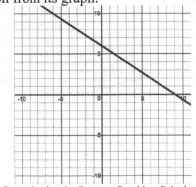
Created using the Desmos Graphing Calculator	Created using the Desmos Graphing Calculator

SOLUTION:

The x-intercept is $(2,0)$ and the y-intercept is $(0,-4)$.

The slope can be computed using the slope formula and those two points.

$$m = \frac{y_2 - y_1}{x_2 - x_1} = \frac{-4 - 0}{0 - 2} = \frac{-4}{-2} = 2$$

The slope is 2.

Exercise 2

Find the slope and any intercepts for the linear function from its graph.

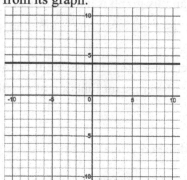

Created using the Desmos Graphing Calculator

Exercise 3

Find the slope and any intercepts for the linear function from its graph.

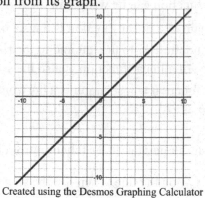

Created using the Desmos Graphing Calculator

Exercise 4

Find the slope and any intercepts for the linear function from its graph.

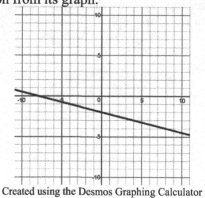

Created using the Desmos Graphing Calculator

Exercise 5

Find the slope and any intercepts for the linear function from its graph.

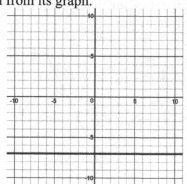

Created using the Desmos Graphing Calculator

6.2.4 Solve applications that involve linear functions as models.

When working with an applied problem that involves a linear function $f(x)$ as a model, you must determine whether you need to evaluate the linear function for a particular value of x or whether you need to set the function equal to a particular value and solve for x.

It helps to list what $f(x)$ represents, as well as what x represents. Then you can examine what information you have to work with, and whether you have a value of x or a value of $f(x)$.

Suppose $f(x) = 75x + 625$ represents the cost (in dollars) for a student who is taking x units (or credits) at your college.
So, x represents the number of units and $f(x)$ represents the cost.

If you were asked to find the cost for a student taking 12 units, the value of x is 12 and the cost $(f(x))$ is unknown. Evaluate the function for $x = 12$.

If you were asked to find the number of units a student who is paying $1375, we know that the cost $f(x)$ is $1375 and the number of units x is unknown. Set the function equal to 1375 and solve for x.

Example	Exercise 1
A cell phone carrier has a family plan whose monthly bill for a family with x lines is given by the linear function $f(x) = 30x + 20$. If a family has a monthly bill of $200, how many lines do they have?	The temperature y in degrees Celsius of a container of water in a room that is 4°C can be approximated by the linear function $f(x) = -0.8x + 76$, where x is the number of minutes after the water was placed in the room. Approximate the temperature of the water 1 hour after it is placed in the room.
SOLUTION:	
In this problem, $f(x)$ represents the monthly cost for a family with x lines.	
Since the family has a monthly bill of $200, set the function $f(x)$ equal to 200 and solve.	
$$f(x) = 200$$ $$30x + 20 = 200$$ $$30x = 180$$ $$x = 6$$	
The family has 6 lines.	

Exercise 2

The temperature y in degrees Celsius of a container of water in a room that is 4°C can be approximated by the linear function $f(x) = -0.8x + 76$, where x is the number of minutes after the water was placed in the room. How long will it take for the water to reach a temperature of 50°C?

Exercise 3

The number of women earning a master's degree in engineering in a particular year can be approximated by the linear function $f(x) = 600x + 8600$, where x is the number of years after 2009. In what year will the number of women who earn a master's degree in engineering reach 20,000?

Exercise 4

The number of U.S. households with a net worth of at least $1 million (excluding their primary residence) can be approximated by the linear function $f(x) = 600,000x + 6,800,000$, where x represents the number of years after 2008. In what year will the number of households with a net worth of $1 million (excluding their primary residence) reach 20,000,000?

Exercise 5

An entrepreneur has started a business selling smartphone cases. The profit earned after selling x cases is given by the function $f(x) = 10x - 75,000$. How many cases must be sold before the business breaks even?

7.1.1 Interpret and draw line graphs.

A **line graph** is often used to show how a quantity changes over time.

The horizontal axis is used to show time, and the vertical axis is used to show the quantity.

Determine the largest value of the quantity in the data, then construct the vertical axis so that the largest value will fit on your graph.

Many line graphs show how a quantity changes by year. Years that are relatively current are numbers that are quite large, and you can start to label the horizontal axis with the first year just a unit to the right of the vertical axis. Once you decide where to start labeling the horizontal axis, you must apply a consistent scale from that point on.

To construct the line graph, plot the first two ordered pairs. Connect those two points with a line segment.

Plot the third ordered pair, and connect it to the second ordered pair with a line segment.

Repeat this process for each successive ordered pair, connecting it to the preceding ordered pair.

Example	Exercise 1
The following table shows the annual tuition at public four-year colleges and universities, adjusted to reflect 2017 dollars, for various years from 1987 through 2017. Draw a line graph showing the change in tuition over time.	The following table shows a college student's cumulative GPA at the end of each of her first 6 semesters. Draw a line graph showing the change in GPA over time.

Example

Year	1987	1997	2007	2017
Tuition	$3190	$4740	$7280	$9970

Source: College Board

SOLUTION:
Plot each ordered pair and connect them.

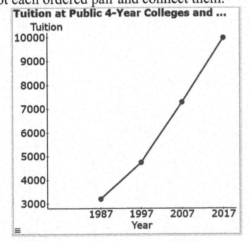

Exercise 1

Semester	1	2	3	4	5	6
GPA	3.5	3.25	3.33	3.41	3.56	3.5

Copyright © 2020 Pearson Education, Inc.

Exercise 2
The following table shows a student's exam scores by exam. Draw a line graph showing the change in exam scores over the semester.

Exam	1	2	3	4	5
Score	84	90	96	92	99

Exercise 3
The following table shows the number of wins by the Boston Red Sox for the seasons from 2013 to 2018. Draw a line graph showing the number of wins by season.

Year	2013	2014	2015	2016	2017	2018
Wins	97	71	78	93	93	108

Source: MLB

Exercise 4
The following table shows the median home sale price (in $1000's) in California from 2005 – 2013. Draw a line graph showing the change in the median home price over time.

Year	2005	2007	2009	2011	2013
Median Price ($1000's)	458	446	328	281	376

Source: Zillow

Exercise 5
The following table shows the number of Starbucks worldwide stores (in hundreds) from 2004 – 2012. Draw a line graph showing the change in the number of stores over time.

Year	2004	2006	2008	2010	2012
Stores (100's)	86	124	167	169	176

Source: Starbucks

7.1.2 Interpret and draw bar graphs.

A **bar graph** is used to show how often a category is present in a set of data.

Bar graphs are useful for comparing the quantity of one category to other categories.

Before constructing a bar graph, you must determine the frequency for each category.

Determine the largest frequency, then construct the vertical axis so that the largest frequency will fit on your graph.

Label the horizontal axis with all of the categories, equally spaced.
For some graphs there will be a natural order to list the categories in, like Winter, Spring, Summer, Fall.
For other graphs you may choose to order the categories in terms of their order in the data set or table.

For each category, draw a bar to the height corresponding to its frequency.

The bars should be of equal widths, and there should be a gap between each pair of bars.

Example	Exercise 1
A sample of 75 college students was made up of 22 freshmen, 20 sophomores, 18 juniors, and 15 seniors. Create a bar graph for this data. SOLUTION: For each group, draw a bar to the desired height. 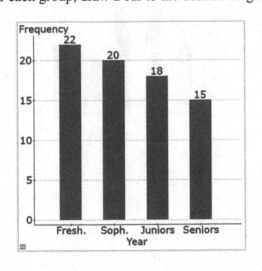	Fifty college students who owned a smartphone were asked what type of phone they owned: iOS, Android, or other. Eighteen owned an iOS phone, 31 owned an Android phone, and 1 listed other. Create a bar graph for this data.

Exercise 2

A community college counselor asked 30 of her students to tell her which school they would most likely transfer. Create a bar graph for this data.

School	Count
Fresno State	15
UC Davis	8
Cal Poly	4
Other	3

Exercise 3

A lab randomly selected 20 blood samples and recorded the patient's blood type. Create a bar graph for this data.

O A A A O
A O A O O
O B O O A
B A O A AB

Exercise 4

The bar graph shows the performance on a math exam by letter grade. Which letter grade was earned by the greatest number of students?

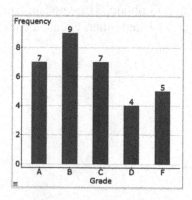

Exercise 5

The bar graph shows the performance on a math exam by letter grade. How many more students earned a C than a D?

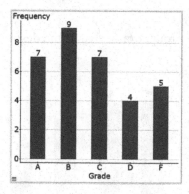

7.1.3   Construct frequency distributions and relative frequency distributions for a data set.

A **frequency distribution** can be used to present numerical or categorical data in a table.

For categorical data, in the first column list all of the different categories in the first column.
In the next column, list the frequency for each category.

For numerical data, divide the data into **classes**, which are ranges of values.
For example, when making a frequency distribution for test scores you might choose the following classes: 40 to 49, 50 to 59, 60 to 69, …
Choose your classes so each value in the data set fits in one of the classes and there are no gaps between classes,
The first number listed is called the lower class limit.

Find the frequency for each class.
List the classes in the first column, and their frequencies in the second column.

To create a **relative frequency distribution** from either type of frequency distribution, find the relative frequency for each category or class by dividing its frequency by the total of all the frequencies.

Example	Exercise 1
A math instructor gave an exam to three classes. Here is the combined grade breakdown for the three classes. A: 20, B: 32, C: 18, D: 8, F:2 Create a relative frequency distribution for the letter grades. SOLUTION: There are 80 grades. To find the relative frequency for each grade, divide the number of students who earned that grade by 80.	Two hundred students attended a guest lecture. There were 68 freshmen, 52 sophomores, 35 juniors, 30 seniors, and 15 grad students. Create a relative frequency distribution for the data.

Grade	Rel. Freq.
A	0.25
B	0.40
C	0.225
D	0.1
F	0.025

Exercise 2
A lab randomly selected 20 blood samples and recorded the patient's blood type. Create a frequency distribution for the blood types.

O A A A O
A O A O O
O B O O A
B A O A AB

Exercise 3
A lab randomly selected 20 blood samples and recorded the patient's blood type. Create a relative frequency distribution for the blood types.

O A A A O
A O A O O
O B O O A
B A O A AB

Exercise 4
Create a frequency distribution for the ages of the U.S. presidents (Washington through Trump) at inauguration.

57	61	57	57	58	57	61	54	68
51	49	64	50	48	65	52	56	46
54	49	51	47	55	55	54	42	51
56	55	51	54	51	60	60	43	55
56	61	52	69	64	46	54	47	70

Age	Frequency
40 to 49	
50 to 59	
60 to 69	
70 to 79	

Exercise 5
Create a relative frequency distribution for the ages of the U.S. presidents (Washington through Trump) at inauguration.
(Round relative frequencies to the nearest thousandth.)

57	61	57	57	58	57	61	54	68
51	49	64	50	48	65	52	56	46
54	49	51	47	55	55	54	42	51
56	55	51	54	51	60	60	43	55
56	61	52	69	64	46	54	47	70

Age	Relative Frequency
40 to 49	
50 to 59	
60 to 69	
70 to 79	

7.1.4 Interpret and draw histograms.

A **histogram** is a graphical representation of a frequency distribution for numerical data.

It is similar to a bar graph, with two exceptions.

- Because the data are numerical, the horizontal axis will be labeled with a numerical scale.

- The bars for each class will have no gaps between them.

Plot each lower class limit on the horizontal axis.

You should also plot the value that would be the lower class limit of the next class.

For example, if the five classes were 0 to 19, 20 to 39, 40 to 59, 60 to 79, and 80 to 100, plot the lower class limits (0, 20, 40, 60, and 80) as well as 100.

Draw the first bar to the height of its frequency above the first class.
The bar should extend from the first lower class limit to the next lower class limit.

Repeat for each class, with no gaps between the bars.

Example	**Exercise 1**
Create a histogram from the frequency distribution of test scores.	Create a histogram from the frequency distribution of ages of math instructors.

Example

Test Score	Frequency
50 to 59	7
60 to 69	3
70 to 79	12
80 to 99	18
90 to 100	10

SOLUTION:
Make marks on the horizontal axis for 50, 60, 70, 80, 90, and 100. Draw bars for each class to the height of its frequency.

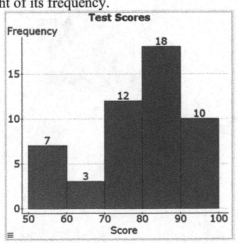

Exercise 1

Age	Frequency
25 to 34	2
35 to 44	9
45 to 54	11
55 to 64	9

Exercise 2
Create a histogram from the frequency distribution of ages of US presidents at inauguration.

Age	Frequency
40 to 49	9
50 to 59	25
60 to 69	10
70 to 79	1

Exercise 3
Create a histogram from the frequency distribution of team home runs in 2018 by MLB teams.

Home Runs	Frequency
120 to 149	3
150 to 179	11
180 to 209	8
210 to 239	7
240 to 269	1

Exercise 4
The following histogram shows the breakdown of exam scores in a math class. How many more students scored in the 80's than in the 70's?

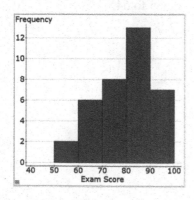

Exercise 5
The following histogram shows the breakdown of exam scores in a math class. If a score of 70 or higher is considered passing, how many students passed the exam?

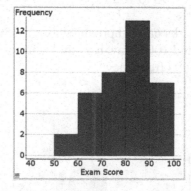

7.1.5 Interpret and draw circle graphs.

A **circle graph**, also known as a **pie chart**, is used with categorical data to show what portion of the total each category represents.

The circle represents 100%, and it will be divided into slices to represent the percent for each category.

To create a circle graph, first determine what percent of the whole each category represents.

This is done by dividing the frequency for each class by the total frequency for all of the classes. Convert to a percent by multiplying by 100%.

Drawing the Graph
- Draw a circle with a point at its center.
- Divide the circle into quarters with a mark on the top and bottom of the circle, as well as on the left and right of the circle. (Where you find 3, 6, 9, and 12 on a clock or watch.)
- Starting at the mark on the right side, label that as 0%.
- Proceeding counter-clockwise, label the other marks as 25%, 50%, and 75%.
- You can add 4 marks in each quarter-circle to denote steps of 5%.
- Draw a line segment from the center of the circle to the mark at 0% and let that be the beginning of your first slice. After you draw the first slice to end at the desired percent, let that be the start of your second slice. Repeat until all slices are complete.

Example	Exercise 1
The students in a statistics class took an exam. Twenty-eight students passed the exam and 7 students failed. Create a circle graph for this data.	A pollster asked a random sample of registered voters if they planned to vote for a certain candidate. Fifty-five percent of those surveyed said yes, 40% said no, and 5% were undecided. Create a circle graph for this data.
SOLUTION: Divide 28 by 35 to determine that 80% of the students passed. Divide 7 by 35 to determine that 20% of the students failed. Draw the slices to represent 80% and 20% of the circle. 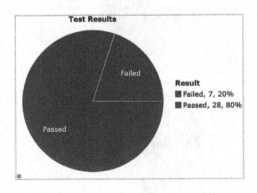	

Exercise 2

A sample of 80 college students was made up of 32 freshmen, 20 sophomores, 16 juniors, and 12 seniors. Create a circle graph for this data.

Exercise 3

Fifty college students who owned a smartphone were asked what type of phone they owned: iOS, Android, or other. Eighteen owned an iOS phone, 31 owned an Android phone, and 1 listed other. Create a circle graph for this data.

Exercise 4

A lab randomly selected 20 blood samples and recorded the patient's blood type. Create a circle graph for this data.

O A A A O
A O A O O
O B O O A
B A O A AB

Exercise 5

A sample of adults was asked for their education status and whether they were a smoker. The results are displayed in the following circle graphs. Based on the graphs, who is more likely to be a smoker – high school graduates or adults with a bachelor's degree?

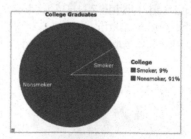

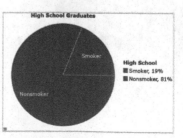

7.1.6 Interpret and build scatterplots.

A **scatterplot** is used to display paired (or bivariate) numerical data.

Create a rectangular grid and label it so that all ordered pairs can be graphed.
(Determine the minimum and maximum values of both x and y, and be sure that your graph extends far enough to cover each of these values.)

Plot each point as you would plot an ordered pair.

Do not connect the points.

A scatterplot can be used to examine whether a relationship exists between two numerical variables.

Example	Exercise 1
A weight is paced on the end of a spring and the displacement (distance that the spring is stretched) is measured. Create a scatterplot for the data.	Exposure to radiation (rem) is measured for certain distances (ft) from the source of radiation. Create a scatterplot for the data.

Weight (g)	50	100	150	200
Displacement (cm)	2	4	6	8

Distance (ft)	6	8	10	12	16	24
Exposure (rem)	142	80	51	36	20	9

SOLUTION:
Plot each of the four ordered pairs.

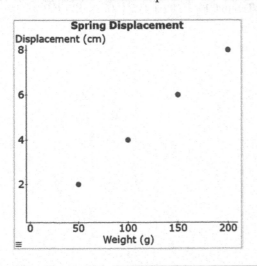

Exercise 2

Number of bacteria cells (in 1000's) counted x minutes after the start of an experiment. Create a scatterplot for the data.

Time (minutes)	10	20	30	45	60	90
Cells (1000's)	7	10	14	24	40	113

Exercise 3

A projectile is launched with an initial velocity of 128 ft/sec at an angle of 30° with the ground. The height of the projectile (ft) is measured at certain time intervals. Create a scatterplot for the data.

Time (sec)	0.5	1	1.5	2	2.5	3	3.5	4
Height (ft)	28	48	60	64	60	48	28	0

Exercise 4

The body of a person who has just died is placed in a refrigerated room that is kept at a constant temperature of 35° F. The body's temperature is recorded at certain intervals. Create a scatterplot for the data.

Time (hr)	1	2	5	10	12	24
Temp. (°F)	98	97	94	89	87	78

Exercise 5

A pot of water is brought to a boil (212° F) and then left in a room that is kept at a constant temperature of 70° F. The water's temperature is recorded at certain intervals. Create a scatterplot for the data.

Time (min)	5	10	20	30	45	60
Temp. (°F)	190	172	143	123	102	89

7.2.1 Find the mean of a data set.

In statistics, **measures of center** are used to describe the center of a data set.

These are used to help describe a typical value for a data set.

Mean
The **mean** of a set of data is a measure of center for a set of data.

To calculate the mean for a set of data, simply find the total of all of the values and divide by the number of values.

The mean is the arithmetic average of the set of data.

$$\text{Mean} = \frac{\text{Sum of All Values}}{\text{Number of Values}}$$

The formula can also be expressed as

$$\text{Mean} = \frac{\Sigma x}{n}$$

The symbol Σ is the capital Greek letter sigma, and is often used in mathematics to denote a sum. Σx represents the sum of all the values, and n represents the number of values in the data set.

Example	Exercise 1
Find the mean for the given values: 63, 98, 21, 42, 71 (Round to the nearest tenth if necessary.)	Find the mean for the given values: 3, 17, 21, 35, 42, 59, 89, 106 (Round to the nearest tenth if necessary.)
SOLUTION: First find the sum of these 5 values. $$63 + 98 + 21 + 42 + 71 = 295$$ To find the mean, divide this sum by 5. $$\text{Mean} = \frac{295}{5} = 59$$ The mean is 59.	

Exercise 2

Find the mean. Round to the nearest tenth if necessary.

IQ of 12 college students:

95	82	104	119	118	126
82	86	116	85	90	90

Exercise 3

Find the mean. Round to the nearest tenth if necessary.

Red Sox home runs (2001-2008):

198, 177, 238, 222, 199, 192, 166, 173

Exercise 4

Find the mean. Round to the nearest tenth if necessary.

Systolic blood pressure (mmHg) of nine 60-year-old women:

119, 160, 121, 92, 109, 95, 114, 112, 122

Exercise 5

Find the mean.

Starting salary for 5 bachelor's degrees:

Chemical Eng.	Computer Science	Mathematics	Political Science	English
$61,800	$54,200	$43,500	$39,400	$36,700

Source: PayScale.com

7.2.2 Find the weighted mean.

The **weighted mean** of a set of data is a measure of center that is used when certain values are repeated in the set of data.

The weighted mean is also used when certain values have greater significance or importance than others, such as when computing grades. If 30% of your overall grade comes from homework and 70% of your overall grade comes from exams, your homework grade has a weight of 0.3 (30%) and your exam grade has a weight of 0.7 (70%).

To calculate the weighted mean,
- Identify the weight w of each value x.
- Multiply each value x by its weight.
- Find the sum of these products.
- Divide the sum by the total of the weights.

Formula for Weighted Mean: $$\text{Weighted Mean} = \frac{\sum(w \cdot x)}{\sum w}$$

Consider the data set 20, 30, 30, 30, 50, 50. The value 20 has a weight of 1 because it only appears once. The value 30 has a weight of 3, and the value 50 has a weight of 2.

$$\text{Weighted Mean} = \frac{\sum(w \cdot x)}{\sum w} = \frac{1 \cdot 20 + 3 \cdot 30 + 2 \cdot 50}{1 + 3 + 2} = \frac{210}{6} = 35$$

Example	Exercise 1
Over a series of 8 basketball games, a player had six personal fouls 1 time, five personal fouls 4 times, four personal fouls 2 times, and two personal fouls 1 time. Use the weighted mean formula to compute the mean number of personal fouls for these 8 games.	On the 5 exams this semester, a student scored 90 twice, 94 twice, and 78 once. Use the weighted mean formula to compute the mean score.

SOLUTION:
The numbers of fouls are the values in this problem, and the weights are the number of times they occurred.

$$\frac{\sum(w_i \cdot x_i)}{\sum w_i} = \frac{1 \cdot 6 + 4 \cdot 5 + 2 \cdot 4 + 1 \cdot 2}{1 + 4 + 2 + 1}$$

$$= \frac{36}{8}$$

$$= 4.5$$

The mean number of personal fouls is 4.5.

Exercise 2

When computing a student's GPA, an A counts for 4 points, a B counts for 3 points, a C counts for 2 points, a D counts for 1 point, and an F counts for 0 points. The GPA is weighted by the number of units per class. Compute a student's GPA for the following grades.
Round to the nearest hundredth.

Class	Units	Grade
Math	4	A
Physics	5	A
Sociology	3	B
English	3	C
US History	4	D

Exercise 3

When computing a student's GPA, an A counts for 4 points, a B counts for 3 points, a C counts for 2 points, a D counts for 1 point, and an F counts for 0 points. The GPA is weighted by the number of units per class. Compute a student's GPA for the following grades.
Round to the nearest hundredth.

Class	Units	Grade
Math	3	A
English	3	A
Philosophy	3	B
Business	5	F

Exercise 4

In a statistics class, homework is weighted at 30%, tests at 50%, and the final exam at 20%. If a student has a homework grade of 84, a test grade of 92, and a final exam score of 98, use the weighted mean formula to compute the student's grade.

Exercise 5

In a math class, homework is weighted at 15%, quizzes at 10%, projects at 40%, the midterm exam at 15%, and the final exam at 20%. Use the weighted mean formula to compute the grade for a student with the following scores.

Homework: 75, Quizzes: 62, Projects: 79, Midterm exam: 84, Final exam: 78

7.2.3   Find the median of a data set.

The **median** of a set of data is another measure of center for a set of data.

The median is often used for types of data that have unusually high or low values, such as home prices and family income.

To find the median for a set of data, begin by writing the values in ascending order.
The median separates the upper half of the values from the lower half.

If a set of data has an odd number of values, the single value in the middle of the data set is the median.

For the following set of seven values, the median is 40.
Notice that three of the values are less than the median, and three of the values are greater than the median.

10   15   22 | 40 | 41   49   60

If a set of data has an even number of values, the median is the mean of the two center values.

For the following set of eight values, the median is 19, which is the mean of the two values in the center: 17 and 21.

6   9   10 | 17   21 | 22   30   70

Example	**Exercise 1**
Find the median for the given values: 63, 98, 21, 42, 71 SOLUTION: Begin by arranging the values in ascending order. 21, 42, 63, 71, 98 The median is the single value in the middle, which is 63.	Find the median for the given values: 3, 17, 21, 35, 42, 59, 89, 106

Exercise 2
Find the median.
Number of Facebook friends for 6 college math instructors:
$$243, 18, 21, 152, 93, 125$$

Exercise 3
Find the median.
Serum glucose level (mg/dL) of 8 people:
$$90, 91, 94, 122, 113, 142, 59, 92$$

Exercise 4
Find the median.
Math test scores of 9 students in a study group:
$$80, 96, 100, 89, 74, 96, 95, 98, 87$$

Exercise 5
Find the median.
Pregnancy duration (days) for 11 women:

267	255	263	261	265	273
264	267	268	275	273	

7.2.4 Find the mode of a data set.

The **mode** of a set of data is the value that is repeated most often.

Consider the data set
$$9, 12, 16, 16, 20, 24, 24, 24, 29, 30$$
The mode is 24, because it is repeated 3 times. No other value is repeated as many times.

If there are no repeated values, the set of data has no mode.

Consider the data set
$$1, 2, 3, 5, 6, 8, 9, 13, 14$$
This data set has no mode because no value is repeated.

A set of data can have more than one mode if two or more values are repeated the same number of times.

Consider the data set
$$2, 5, 5, 7, 8, 8, 9$$
This data set has two modes, 5 and 8. They are both repeated two times, which is more times than any other value.

Example	**Exercise 1**

Find the mode, if it exists. If there is more than one mode, list each mode.
    70, 56, 63, 35, 56, 63, 36, 56, 19
 	
SOLUTION:
The value 56 is repeated three times. Since no other value is repeated that many times, 56 is the mode.
	Find the mode, if it exists. If there is more than one mode, list each mode.
	    7, 41, 32, 56, 41, 19, 8, 32, 25

Exercise 2
Find the mode, if it exists. If there is more than one mode, list each mode.

5, 35, 89, 106, 42, 17, 59, 21

Exercise 3
Find the mode, if it exists. If there is more than one mode, list each mode.
Nightly room rate at 10 Las Vegas hotels for Valentine's Day:

$219	$259	$127	$199	$259
$169	$219	$229	$299	$199

Exercise 4
Find the mode, if it exists. If there is more than one mode, list each mode.
Touchdown passes thrown by Joe Montana by year (16 seasons):

1	15	19	17	26	28	27	8
31	18	26	26	0	2	13	16

Exercise 5
Find the mode, if it exists. If there is more than one mode, list each mode.
Time (seconds) of the 8 songs on Bruce Springsteen's *Born to Run*:

289, 191, 180, 390, 271, 270, 198, 574

7.2.5 Find the range and midrange of a data set.

The **range** is a measure of dispersion for a set of data.

A measure of dispersion measures how "spread out" a set of data is.

The range of a data set is equal to the difference between its largest and smallest values.

$$\text{Range} = \text{Maximum Value} - \text{Minimum Value}$$

To find the range, first identify the data set's maximum and minimum values. Then subtract those two values.

The **midrange** for a set of data is a measure of center that can be quickly calculated.

The midrange is exactly halfway between a data set's minimum and maximum values. It is the mean of those two values.

$$\text{Midrange} = \frac{\text{Minimum Value} + \text{Maximum Value}}{2}$$

Example	Exercise 1
Find the range for the given values: 63, 98, 21, 42, 71 SOLUTION: The maximum value is 98, and the minimum value is 21. To find the range, find the difference between those two values. Range: $98 - 21 = 77$	Find the midrange for the given values: 3, 17, 21, 35, 42, 59, 89, 106

Exercise 2
Find the range.
Mobile phone minutes used by 14 families last month:

636	754	662	884	1346	659	1006
1357	1129	904	1747	1336	1234	388

Exercise 3
Find the midrange.
Mobile phone minutes used by 14 families last month:

636	754	662	884	1346	659	1006
1357	1129	904	1747	1336	1234	388

Exercise 4
Find the range.
Systolic blood pressure (mmHg) of thirteen 65-year-old smokers:

110	118	137	127	134	163	129
102	102	136	150	130	113	

Exercise 5
Find the midrange.
Systolic blood pressure (mmHg) of thirteen 65-year-old smokers:

110	118	137	127	134	163	129
102	102	136	150	130	113	

8.1.1 Evaluate factorial expressions.

The expression $n!$, called **n factorial**, is the product of all integers from n down to 1.
$n!$ can be used to compute the number of different ways that n distinct items can be ordered.

For example, $8! = 8 \cdot 7 \cdot 6 \cdot 5 \cdot 4 \cdot 3 \cdot 2 \cdot 1 = 40{,}320$.

Most scientific calculators have a built-in function for computing factorials.
Consult the manual for your calculator to learn how to compute factorials on your calculator.
Here is a screen shot for computing $8!$ using the Texas Instruments TI-84 calculator.

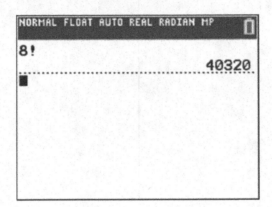

Example	**Exercise 1**
Compute: $6!$	Compute: $3!$
SOLUTION:	
Rewrite $6!$ as a product and simplify.	
$$6! = 6 \cdot 5 \cdot 4 \cdot 3 \cdot 2 \cdot 1 = 720$$	

Exercise 2
Compute: 5!

Exercise 3
Compute: 7!

Exercise 4
Compute: $\dfrac{9!}{3!}$

Exercise 5
Compute: $\dfrac{9!}{4! \cdot 5!}$

8.1.2 Use counting techniques.

Counting techniques are used to determine the number of possible outcomes for a given scenario.

One tool for counting possibilities is the **multiplication principle**.

Multiplication Principle
If one choice can be made in m ways and a second choice can be made in n ways, then the two choices can be made in $m \cdot n$ ways.

For example, suppose that a gym has 5 kickboxing classes and 8 spinning classes. If no two classes have a time conflict, how many different ways are there to sign up for 1 kickboxing class and 1 spinning class?

To answer this question, multiply the number of kickboxing classes by the number of spinning classes.

$$5 \cdot 8 = 40 \text{ different ways}$$

The multiplication principle also applies when there are more than two choices to make.

Example	Exercise 1
A restaurant has 7 types of dessert and 4 types of coffee. In how many different ways can you order dessert and coffee? SOLUTION: Multiply the number of desserts by the number of types of coffee. $$7 \cdot 4 = 28$$ There are 28 different ways you could order coffee and dessert.	An ice cream shop has 12 flavors of ice cream, 6 toppings, and 4 types of cones. In how many ways could you order an ice cream cone? (1 flavor, 1 topping, 1 cone)

Exercise 2
License plates in Rhode Island have 2 letters followed by 3 digits. How many different license plates are possible?

Exercise 3
Eight runners are in a race. In how many different ways can they finish first, second, and third?

Exercise 4
A password is made up of 5 digits. How many different passwords are possible if no digit can be used more than once?

Exercise 5
There are 20 students in a classroom. In how many different ways can an instructor pick one student to get an A, a second student to get a B, a third student to get a C, and a fourth student to get an F?

8.2.1 Identify sample spaces, outcomes, and events.

An **outcome** is a result of a probability experiment.

For example, if a single six-sided die is rolled and the result is a 5, then 5 is an outcome of the experiment.

The set of all possible outcomes of a probability experiment is called the **sample space**.
The sample space is often denoted as S.

Again, if a single six-sided die is rolled, the possible outcomes are 1, 2, 3, 4, 5, and 6.
So, $S = \{1,2,3,4,5,6\}$.

An **event** is some collection of outcomes. An event can be made up of multiple outcomes, a single outcome, or even zero outcomes.
Events are typically denoted by capital letters, other than S.

If a single six-sided die is rolled, some events include:
- The result is odd: $A = \{1,3,5\}$
- The result is less than 5: $B = \{1,2,3,4\}$
- The result is positive: $C = \{1,2,3,4,5,6\}$

Example	**Exercise 1**
A fair coin is flipped two times. Each time either heads or tails is recorded. List the sample space for this experiment. SOLUTION: Let H represent a flip that is heads and T represent a flip that is tails. Sample space: $\{HH, HT, TH, TT\}$	A fair coin is flipped two times. Each time either heads or tails is recorded. List the outcomes that are in the event A: There is exactly one head in the two flips.

Exercise 2

A single six-sided die is rolled. List the outcomes in the sample space S.

Exercise 3

A single six-sided die is rolled. List the possible outcomes in the event A: The roll is odd.

Exercise 4

A pair of fair 6-sided dice are rolled and the sum is recorded. Use the listed sample space of 36 possible rolls to determine the number of possible outcomes for the event A: The sum is 8 or higher.

SUM

	Second Die					
	1	**2**	**3**	**4**	**5**	**6**
1	2	3	4	5	6	7
2	3	4	5	6	7	8
First Die 3	4	5	6	7	8	9
4	5	6	7	8	9	10
5	6	7	8	9	10	11
6	7	8	9	10	11	12

Exercise 5

A pair of fair 6-sided dice are rolled and the sum is recorded. Use the listed sample space of 36 possible rolls to determine the number of possible outcomes for the event A: The sum is 10 or lower.

SUM

	Second Die					
	1	**2**	**3**	**4**	**5**	**6**
1	2	3	4	5	6	7
2	3	4	5	6	7	8
First Die 3	4	5	6	7	8	9
4	5	6	7	8	9	10
5	6	7	8	9	10	11
6	7	8	9	10	11	12

8.2.2 Find the probability of an event.

The probability that an event occurs is a measure of the likelihood of that event occurring.
The probability of an event occurring is a number between 0 (0%) and 1 (100%).

If all of the possible outcomes of a probability experiment are equally likely to occur, then the
classical method can be used to compute the probability of an event A occurring.

$$P(A) = \frac{n(A)}{n(S)}$$

$P(A)$ represents the probability of event A occurring, $n(A)$ is the number of outcomes in event A,
and $n(S)$ is the number of outcomes in the sample space, S.

For example, in a standard 52-card deck of playing cards, there are 4 aces.
The deck is shuffled, and a card is selected at random.
Let A be the event that the card is an ace. Since there are 4 aces in the deck, $n(A) = 4$. The sample
space consists of the 52 different cards, so $n(S) = 52$.

$$P(A) = \frac{n(A)}{n(S)} = \frac{4}{52} = \frac{1}{13}$$

Be sure to simplify the fraction to lowest terms.

Example	Exercise 1
A fair 6-sided die is rolled. Find the probability that the roll is even. Write your answer as a fraction in simplest terms.	A fair 6-sided die is rolled. Find the probability that the roll is 5 or less. Write your answer as a fraction in simplest terms.
SOLUTION:	
There are 6 possible rolls: $S = \{1, 2, 3, 4, 5, 6\}$	
Three of the rolls are even: $A = \{2, 4, 6\}$	
The probability that the roll is even is $\frac{3}{6}$, which simplifies to be $\frac{1}{2}$.	

Exercise 2

A statistics class has 30 students. Twenty of the students are female. If a student is selected at random, find the probability that the student is male. Write your answer as a fraction in simplest terms.

Exercise 3

A jar contains 50 red marbles, 20 green marbles, and 10 blue marbles. If a marble is selected at random, find the probability that the marble is green. Write your answer as a fraction in simplest terms.

Exercise 4

A jar contains 60 red marbles, 30 green marbles, and 10 blue marbles. If a marble is selected at random, find the probability that the marble is not green. Write your answer as a fraction in simplest terms.

Exercise 5

A pair of fair 6-sided dice are rolled and the sum is recorded. Use the listed sample space of 36 possible rolls to find the probability that the sum is 9 or lower. Write your answer as a fraction in simplest terms.

SUM		Second Die					
		1	**2**	**3**	**4**	**5**	**6**
	1	2	3	4	5	6	7
	2	3	4	5	6	7	8
First Die	**3**	4	5	6	7	8	9
	4	5	6	7	8	9	10
	5	6	7	8	9	10	11
	6	7	8	9	10	11	12

8.2.3   Use tree diagrams to find sample spaces and compute probabilities.

A **tree diagram** is a useful tool for determining the sample space (S) for an experiment by listing out all of the possible outcomes of a probability experiment.

Suppose a coin is flipped and the result (heads or tails) is recorded.
The coin is then flipped a second time and the result of this second trial is recorded.
On the first level of the tree, list the two possible outcomes: H and T.
These represent the possible results for the first toss.
If the first toss is heads, list the possible outcomes of the second toss on branches leading to the second level. Do the same for the branch of the tree with tails on the first toss.
Here is an example of how this should look.

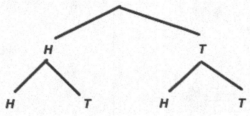

Looking to the second level, there are a total of four branches. From left to right, those four branches are HH, HT, TH, TT. Those four outcomes make up the sample space.

The probability that both tosses are tails can be found from the tree diagram. Only 1 of the 4 possible outcomes contains two tails, so $P(2\ T) = \dfrac{1}{4}$.

Example	**Exercise 1**
A fair coin is flipped three times. Each flip is recorded as heads or tails. Use a tree diagram to determine the number of possible outcomes in the sample space.	A fair coin is flipped three times. Each flip is recorded as heads or tails. Use a tree diagram to find the probability that exactly two of the flips are heads. Write your answer as a fraction in simplest terms.

SOLUTION:
There are 3 levels to this experiment, one for each flip of the coin. Each flip of the coin can be either heads (H) or tails (T).

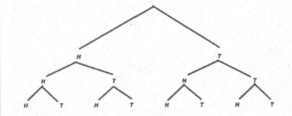

There are 8 possible outcomes.

Exercise 2

A fair coin is flipped three times. Each flip is recorded as heads or tails. Use a tree diagram to find the probability that at least one of the flips are heads. Write your answer as a fraction in simplest terms.

Exercise 3

A couple has four children. Use a tree diagram to determine the number of possible outcomes in the sample space with respect to the gender of each child.

Exercise 4

A couple has four children. Use a tree diagram to find the probability that the couple has exactly two girls. Write your answer as a fraction in simplest terms.

Exercise 5

A couple has four children. Use a tree diagram to find the probability that the couple has more girls than boys. Write your answer as a fraction in simplest terms.

9.1.1 Define terminology associated with sets.

A **set** is a collection of objects.

In mathematics we typically deal with sets of numbers.

The members of a set are called **elements**.

The elements of a set are typically written inside a set of curly brackets.

The set of possible rolls of a 6-sided die are the whole numbers from 1 through 6: $\{1, 2, 3, 4, 5, 6\}$.

A set A is a subset of another set B if every element of A is also an element of B.

For example, the set of college freshmen is a subset of the set of college students because every college freshman is a college student.

The **empty set** is a set that has no elements.

An empty set can be written as $\{\ \}$ or $\varnothing$.

Example	**Exercise 1**
Identify the term being identified.	Identify the term being identified.
A collection of objects.	A member of a set.
SOLUTION:	
This is the definition of a set.	
A set is a collection of objects.	

Exercise 2 Identify the term being identified. A set with no elements.	**Exercise 3** List the condition for a set *A* to be a subset of another set *B*.
Exercise 4 True or False: A set is always a subset of itself.	**Exercise 5** True or False: The empty set is a subset of every set.

9.1.2 Describe the members of a set using various notations.

There are several notations that can be used to describe a set.

Roster Notation
To describe the members of a set using roster notation, list the members inside a set of curly brackets, separated by commas.

The set of months of the year can be expressed using roster notation as
$$\{\text{January, February, March, ..., November, December}\}$$

Interval Notation
A set of real numbers can be expressed using interval notation, such as the solutions of an inequality.

The set of all real numbers that are less than or equal to 5 can be expressed in interval notation as $(-\infty, 5]$.

Set-Builder Notation
A set can be expressed using set-builder notation by listing the definition of the members of that set.

The set of all real numbers that are less than or equal to 5 can be expressed in set-builder notation as $\{x \mid x \leq 5\}$.

Example	Exercise 1
Describe the elements of the set using roster notation: Months of the year	Describe the elements of the set using interval notation: Real numbers less than or equal to −2
SOLUTION: Using the roster method, this set is: {January, February, March, April, May, June, July, August, September, October, November, December}	

Exercise 2	**Exercise 3**
Describe the elements of the set using set-builder notation: Real numbers greater than 5	Describe the elements of the set using roster notation: Integers between −4 and 3, inclusive

Exercise 4	**Exercise 5**
Describe the elements of the set using interval notation: Real numbers that are greater than or equal to 10	Describe the elements of the set using set-builder notation: Positive real numbers

9.1.3   Find subsets of a set.

A set A is a subset of the set B if every element (or member) of A is also an element (or member) of B.

To show that a set A is not a subset of another set B, you must find an element of A that is not an element of B.

For example, suppose A is the set of prime numbers and B is the set of odd numbers.
To show that A is not a subset of B, you must find a prime number that is not an odd number.
Since 2 is a prime number but not an odd number, then A is not a subset of B.

The empty set is always a subset of any other set because there are no members of the empty set that are not members of the other set.

Suppose you have the set $A = \{1, 2, 3, 4\}$.
There are six different subsets of A that have two elements:
$$\{1, 2\}, \{1, 3\}, \{1, 4\}, \{2, 3\}, \{2, 4\} \text{ and } \{3, 4\}.$$

Example	Exercise 1
Let $A = \{1, 2, 3, 4, 5, 6, 7, 8, 9, 10\}$. List the subset B whose members are members of A that are multiples of 3. SOLUTION: The elements of A that are multiples of 3 are 3, 6, and 9. $$B = \{3, 6, 9\}$$	Let $A = \{1, 2, 3, 4, 5, 6, 7, 8, 9, 10\}$. List the subset B whose members are members of A that are less than 8.

Exercise 2

Let $A = \{1,2,3,4,5,6,7,8,9,10\}$. List the subset B whose members are members of A that are prime numbers.

Exercise 3

Let A be the set of months of the year. List the subset B of months that end with the letters "ber."

Exercise 4

Let A be the set of months of the year. List the subset B of months that start with the letter J.

Exercise 5

Let A be the set of the first 8 prime numbers. List the subset B whose members are members of A that are odd.

9.1.4 Find equivalent sets.

Two sets A and B are **equivalent sets** if A and B have the same number of elements.

If $n(A)$ is the number of elements in set A and $n(B)$ is the number of elements in set B, then sets A and B are equivalent if $n(A) = n(B)$.

Let A be the set of all U.S. states that begin with the letter A:
$$A = \{\text{Alabama, Alaska, Arizona, Arkansas}\}$$

Let B be the set of vowels:
$$B = \{\text{a, e, i, o, u}\}$$

A and B are not equivalent sets because they do not have the same number of elements. $n(A) = 4$, but $n(B) = 5$.

Consider the set C, whose elements are the months that end with the letters "ber".
$$C = \{\text{September, October, November, December}\}$$

Since $n(C) = 4$, the set C is equivalent to the set A, whose elements are the U.S. states that begin with the letter A.

Example	Exercise 1
Are the two sets equivalent? A is the set of days of the week and B is the set of positive integers that are less than 8. SOLUTION: There are 7 days of the week. $A = \{\text{Sunday, Monday, Tuesday, Wednesday, Thursday, Friday, Saturday}\}$ There are 7 positive integers. $B = \{1, 2, 3, 4, 5, 6, 7\}$ Since each set has 7 elements, the two sets are equivalent.	Are the two sets equivalent? A is the set of months of the year and B is the set of possible sums when two 6-sided dice are rolled.

Exercise 2

Are the two sets equivalent?

$A = \{-2, -1, 0, 1, 2\}$ and B is the set of values obtained when the elements of A are squared.

Exercise 3

Are the two sets equivalent?

$A = \{-2, -1, 0, 1, 2\}$ and B is the set of values obtained when the elements of A are multiplied by 2.

Exercise 4

Are the two sets equivalent?

A is the set of different levels of students and B is the set of the number of students in each level.

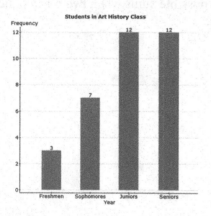

Exercise 5

Are the two sets equivalent?

A is the set of states in the US and B is the set of cards in a standard deck of playing cards.

9.1.5 Find the union of two sets.

The **union** of two sets A and B, denoted $A \cup B$, is the set whose elements are elements of A or B or both.

In other words, the elements of $A \cup B$ are those which are elements of A and/or B.

Every element of A is also an element of $A \cup B$, and the same is true for every element of B.

To list the elements of $A \cup B$, begin with all of the elements of A.
Then list any elements of B that are not also members of A.

For example, suppose $A = \{1, 2, 3, 4, 5\}$ and $B = \{3, 4, 5, 6, 7\}$.
The union of A and B would contain all of the elements of A (1, 2, 3, 4, and 5), as well as the elements of B that are not elements of A (6 and 7).
So, $A \cup B = \{1, 2, 3, 4, 5, 6, 7\}$.

Example	Exercise 1
If A is the set of all whole numbers less than 10 that are divisible by 2, and B is the set of all whole numbers less than 10 that are divisible by 3, find $A \cup B$.	Let $A = \{1, 4, 9, 16, 25, 64\}$ and $B = \{1, 8, 27, 64, 125\}$. Find $A \cup B$.
SOLUTION: Set A is the set of all whole numbers less than 10 that are divisible by 2. So, $A = \{2, 4, 6, 8\}$.	
Set B is the set of all whole numbers less than 10 that are divisible by 3. So, $B = \{3, 6, 9\}$.	
The elements of B that are not elements of A are 3 and 9, so $A \cup B$ contains the elements of A as well as 3 and 9. $$A \cup B = \{2, 3, 4, 6, 8, 9\}$$	

Exercise 2

If A is the set of all people taller than 6 feet tall and B is the set of all women, describe $A \cup B$.

Exercise 3

If A is the set of all numbers less than 5 and B is the set of all numbers greater than 0, describe $A \cup B$.

Exercise 4

Let A be the set of all negative integers and B be the set of all integers that are greater than or equal to 5. Which integers are not elements of $A \cup B$?

Exercise 5

Under what conditions would the number of elements of A plus the number of elements of B be equal to the number of elements in $A \cup B$?

9.1.6 Find the intersection of two sets.

The **intersection** of two sets A and B, denoted $A \cap B$, is the set whose elements are elements of both A and B.

In other words, the elements of $A \cap B$ are the elements which A and B have in common.

Every element of $A \cap B$ is an element of A and also an element of B.

One way to list the elements of $A \cap B$ is to begin with all of the elements of A and then eliminate the elements that are not also elements of B.

For example, suppose $A = \{1, 2, 3, 4, 5\}$ and $B = \{3, 4, 5, 6, 7\}$.

The intersection of A and B would contain all of the elements that A and B have in common, which in this case are 3, 4, and 5.

So, $A \cap B = \{3, 4, 5\}$.

Example	Exercise 1
If A is the set of all whole numbers less than 10 that are divisible by 2, and B is the set of all whole numbers less than 10 that are divisible by 3, find $A \cap B$.	Let $A = \{1, 4, 9, 16, 25, 64\}$ and $B = \{1, 8, 27, 64, 125\}$. Find $A \cap B$.
SOLUTION: Set A is the set of all whole numbers less than 10 that are divisible by 2. So, $A = \{2, 4, 6, 8\}$.	
Set B is the set of all whole numbers less than 10 that are divisible by 3. So, $B = \{3, 6, 9\}$.	
The only element that A and B have in common is 6. $$A \cap B = \{6\}$$	

Exercise 2

If A is the set of all people taller than 6 feet tall and B is the set of all women, describe $A \cap B$.

Exercise 3

If A is the set of all numbers less than 5 and B is the set of all numbers greater than 0, describe $A \cap B$.

Exercise 4

Let A be the set of all numbers greater than -4 and B be the set of all numbers that are less than or equal to 4. List the intersection $A \cap B$ using interval notation.

Exercise 5

Under what conditions would the number of elements of A be equal to the number of elements in $A \cap B$?

9.1.7 Find the complement of a set.

The **universal set**, denoted by U, is the set that contains all elements for a given situation.

For example, if you are working on a problem involving students at your school, the universal set U is the set of all students at your school.

The **complement** of a set A, denoted A', is the set of all elements of the universal set that are not elements of A.

If the universal set U is the set of all students at your school and A is the set of all first-year students at your school, then A' is the set of all students at your school that are not first-year students.

Let $U = \{1,2,3,4,5,6,7,8\}$ and $A = \{3,4,5\}$.

The elements of A' are the members of U other than 3, 4, or 5.

$$A' = \{1,2,6,7,8\}$$

Example	Exercise 1
Let U be the set of all students at a community college, and let A be the set of those students who plan to transfer to a four-year college. Describe the students who are in the complement of set A.	Let U be the set of all items on the menu of a restaurant, and let A be the set of those items that are gluten free. Describe the items that are in the complement of set A.
SOLUTION: The complement of set A is the set of students at a community college that do not plan to transfer to a four-year university.	

Exercise 2

Let U be the set of all real numbers, and let A be the set of real numbers that are greater than 3. Describe the elements of A'.

Exercise 3

Let U be the set of all months, and let A be the set of all months that start with the letter J. List the elements of A'.

Exercise 4

Let $U = \{1, 2, 3, 4, 5, 6, 7, 8, 9, 10\}$, and let $A = \{2, 4, 6, 8, 10\}$. List the elements of A'.

Exercise 5

Let $U = \{1, 2, 3, 4, 5, 6, 7, 8, 9, 10\}$, and let A be the set containing the elements of U that are prime numbers. List the elements of A'.

9.2.1 Distinguish between inductive and deductive reasoning.

Two types of reasoning are inductive reasoning and deductive reasoning.

Inductive reasoning involves moving from a specific statement to a more general statement.

The conclusion of inductive reasoning follows with some likelihood, but not with 100% certainty.

For example, suppose the Red Sox have won their first two games against the Yankees.
Concluding that the Red Sox will beat the Yankees in their third game is an example of inductive reasoning.
Although it seems likely that the Red Sox will beat the Yankees in their third game, that is not guaranteed.

Deductive reasoning involves moving from a general statement to a more specific statement through a logical process.

Unlike inductive reasoning, the conclusion of deductive reasoning follows with 100% certainty.

For example, we know that all islands are surrounded by water, and Oahu is an island.
Concluding that Oahu is surrounded by water is an example of deductive reasoning, and that conclusion is a valid conclusion.

Example	Exercise 1
Is this an example of inductive or deductive reasoning? All humans have brains. Isabella is a human. Therefore, Isabella has a brain. SOLUTION: This is an example of deductive reasoning, because it moves from a general statement ("All humans have brains.") to a more specific statement ("Isabella has a brain.").	Is this an example of inductive or deductive reasoning? I passed the first three exams in my math class. Therefore, I will probably pass the next exam in my math class.

Exercise 2

Is this an example of inductive or deductive reasoning?

All of the students in each of my classes is at least 18 years old. Russell is a student at my college. Therefore, Russell is probably at least 18 years old.

Exercise 3

Is this an example of inductive or deductive reasoning?

Oranges are considered to be citrus fruits. A blood orange is a variety of an orange. Therefore, a blood orange is a citrus fruit.

Exercise 4

Is this an example of inductive or deductive reasoning?

All even numbers are divisible by 2. Twenty is an even number. Therefore, 20 is divisible by 2.

Exercise 5

Is this an example of inductive or deductive reasoning?

Crows, pigeons, blue jays, and hawks are all birds. Crows, pigeons, blue jays, and hawks can all fly. Therefore, all birds can fly.

9.2.2 Distinguish between valid and invalid arguments.

An argument is **valid** if its conclusion necessarily follows from its premises.

Consider the following argument:
- All fish live in water.
- A tuna is a type of fish.

- Therefore, tuna live in water.

This argument is valid, because all fish (including tuna) live in water.

An argument is **invalid** if its conclusion is not a certainty.

Consider the following argument:
- Cola is a beverage.
- Cola is sweet.

- Therefore, all beverages are sweet.

To show that this argument is invalid, you must find a beverage that is not sweet.
Since water is a beverage that is not sweet, the argument is invalid.

Example	**Exercise 1**
Determine whether the argument is valid or not. All vegetarians eat several servings of fruits and vegetables each day. Tina eats several servings of fruits and vegetables each day. Therefore, Tina is a vegetarian.	Determine whether the argument is valid or not. All vegetarians eat several servings of fruits and vegetables each day. Tina is a vegetarian. Therefore, Tina eats several servings of fruits and vegetables each day.
SOLUTION: This argument is not valid because it is possible to eat several servings of fruits and vegetables each day without being a vegetarian.	

Exercise 2	Exercise 3
Determine whether the argument is valid or not. College math instructors have an advanced degree in mathematics. Liana is a college math instructor. Therefore, Liana has an advanced mathematics degree.	Determine whether the argument is valid or not. College math instructors have an advanced degree in mathematics. Liana has an advanced degree in mathematics. Therefore, Liana is a college math instructor.
Exercise 4 Determine whether the argument is valid or not. All the students on the Dean's List study at least 15 hours per week. Fabiola studies at least 15 hours per week. Therefore, Fabiola is on the Dean's List.	**Exercise 5** Determine whether the argument is valid or not. All the students on the Dean's List study at least 15 hours per week. Fabiola is on the Dean's List. Therefore, Fabiola studies at least 15 hours per week.

9.2.3 Identify types of statements.

A logic **statement** is a sentence that is either true or false.

A **quantified statement** is a statement containing a quantifier like none, some, or all, such as "All students like pizza."

The **negation** of a statement is the opposite (or complement) of the statement.
If a statement is true, its negation is false. If a statement is false, its negation is true.

Statements that are formed from two or more statements are called **compound statements**.
Suppose that p and q are statements.
- The compound statement "p and q" is only true if both p and q are true.
- The compound statement "p or q" is true if either p or q (or both) are true.
- The conditional compound statement "if p then q" is true unless p is true but q is false.
- The compound biconditional statement "p if and only if q" means that if either statement is true, then the other statement is true. This type of compound statement is only false if one statement is true but the other is false.

Consider the conditional compound statement "if p then q."
- The **converse** of the statement reverses the order of p and q: "if q then p."
- The **inverse** of the statement negates both p and q, and reverses their order: "if not q then not p."

Example	Exercise 1
True/False: The negation of "It is raining outside" is "It is sunny outside."	Identify the type of statement for the statement "Deborah is majoring in mathematics or she is majoring in physics."
SOLUTION: The negation of the statement is "It is not raining outside." Since this includes options other than "It is sunny outside," such as it is snowing or it is foggy, this is false.	Quantified statement Compound statement

Exercise 2
Identify the type of statement for the statement "All students love math."

Quantified statement
Compound statement

Exercise 3
Identify the type of statement for the statement "The card is an Ace and it is red."

Quantified statement
Compound statement

Exercise 4
Identify the type of statement for the statement "If a positive integer is divisible by 6, then it is divisible by 2."

Quantified statement
Compound statement

Exercise 5
Consider the statement "If a child is eating an apple, then the child is eating fruit." Is the statement "If a child is not eating an apple, then the child is not eating fruit" the converse or inverse of the original statement?

Equivalent Fractions, Adding and Subtracting Fractions

Fractions are used to represent parts of a whole.

1. Match each fraction with the picture whose shaded part represents that fraction.

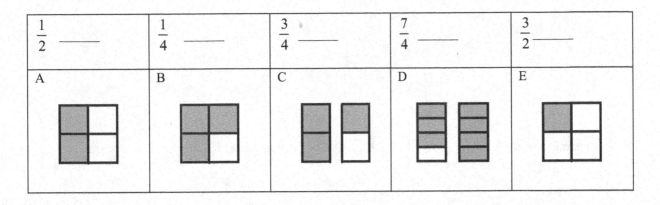

2. Was the answer for $\dfrac{1}{2}$ what you were expecting at first? List another way to express the fraction that matches the picture? _____

3. Which of the following are equivalent to $\dfrac{1}{2}$?

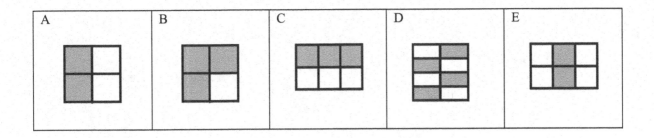

4. Match each fraction to the picture above whose shaded portion is equivalent to that fraction:

$\dfrac{2}{4}$ _____ $\dfrac{4}{8}$ _____ $\dfrac{3}{6}$ _____ $\dfrac{2}{6}$ _____ $\dfrac{3}{4}$ _____

5. Draw a picture whose shaded portion is equivalent to the given fraction and has the stated denominator.

a. $\dfrac{1}{2}$, denominator: 10

b. $\dfrac{3}{4}$, denominator: 8

c. $\dfrac{5}{2}$, denominator: 6

d. $\dfrac{4}{5}$, denominator: 15

e. $\dfrac{2}{10}$, denominator: 5

f. $\dfrac{4}{6}$, denominator: 9

6. Draw a picture that shows that the improper fraction $\dfrac{3}{2}$ is equivalent to the mixed number $1\dfrac{1}{2}$.

Adding and Subtracting Fractions

7. Plot the two fractions on the number line. Find the sum of the two fractions and plot it on the same number line.

a. $\dfrac{1}{2}, \dfrac{1}{3}, \dfrac{1}{2} + \dfrac{1}{3}$

b. $\dfrac{2}{3}, \dfrac{5}{6}, \dfrac{2}{3} + \dfrac{5}{6}$

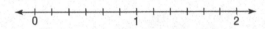

c. $\dfrac{1}{2}, \dfrac{3}{5}, \dfrac{1}{2} + \dfrac{3}{5}$

8. Explain how the sums in exercise 7 could be found from using the number line.

9. Use a number line to find the sum $\dfrac{3}{4} + \dfrac{5}{6}$.

10. Shade the diagram so that the shaded portion is equivalent to the given fraction.

Find the sum $\dfrac{1}{2}+\dfrac{1}{3}$, and shade the diagram so that the shaded portion is equivalent to the sum.

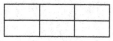

Explain how the diagrams for $\dfrac{1}{2}$ and $\dfrac{1}{3}$ can be used to create a diagram that represents the sum $\dfrac{1}{2}+\dfrac{1}{3}$.

11. Use diagrams to represent the sum $\dfrac{3}{4}+\dfrac{1}{6}$.

$$\frac{3}{4} \qquad\qquad\qquad \frac{1}{6} \qquad\qquad\qquad \frac{3}{4}+\frac{1}{6}$$

Working with Decimals and Fractions

Notice the following matching of a decimal with the corresponding fraction and number word.

Decimal	Fraction	In Words
0.2	$\dfrac{2}{10}$	Two tenths
0.05	$\dfrac{5}{100}$	Five hundredths
0.008	$\dfrac{8}{1000}$	Eight thousandths

1. Complete the following table:

Decimal	Fraction	In Words
a. 0.1		
b. 0.01		
c. 0.001		
d. 0.3		
e. 0.6		
f. 0.8		
g. 0.07		
h. 0.009		
i. 0.020		
j. 0.011		

2. Perform the following operations:

$$7 \cdot \frac{5}{2} \qquad\qquad 7 \cdot 2.5 \qquad\qquad 7 \div \frac{2}{5} \qquad\qquad 7 \div 0.4$$

3. Explain why each of the last three problems in exercise 2 are equivalent to $7 \cdot \dfrac{5}{2}$.

4. Compute the following:

$$1.5^2 \qquad\qquad \left(\frac{3}{2}\right)^2 \qquad\qquad 2.25^3 \qquad\qquad \left(\frac{9}{4}\right)^3$$

5. Explain which pairs of answers in exercise 4 are equal, and why.

6. Compute the following:

$$\left(1.4 + 2.5\right)^3 \qquad\qquad\qquad 1.4^3 + 2.5^3$$

7. Explain why the above two expressions in exercise 6 do NOT give the same answer.

Percent Change

If the price of an item increases by $5, is that a lot? The answer depends upon what the original price was. If the original price was $1 or $1000, the answer is different. To help us understand the size of the change, we can use the idea of percent change.

$$\text{Amount of Increase} = \text{Percent Increase} \cdot \text{Original Value}$$

Find the percent increase if the price increased from $1 to $6.

The amount of increase is $5. $(\$6 - \$1 = \$5)$ Let x represent the percent increase.

$$5 = x \cdot 1$$
$$5 = x$$

Rewriting 5 as a percent, the percent increase is 500%.

Find the percent increase if the price increased from $1000 to $1005.

The amount of increase is $5. $(\$1005 - \$1000 = \$5)$ Let x represent the percent increase.

$$5 = x \cdot 1000$$
$$\frac{5}{1000} = \frac{1000x}{1000}$$
$$0.005 = x$$

Rewriting 0.005 as a percent, the percent increase is 0.5%.
So, we see that the same increase ($5) produces a different percent increase, depending on the original value.

1. Find each of the following percent increases:

	a.	b.	c.	d.
Original Price	$2	$2	$2	$10
New Price	$3	$4	$10	$12
Percent Increase				

A similar process can be used if the price of an item decreases.

$$\text{Amount of Decrease} = \text{Percent Decrease} \cdot \text{Original Value}$$

2. Find each of the following percent decreases. Round to the nearest tenth of a percent.

	a.	b.	c.	d.
Original Price	$6	$7	$12	$100
New Price	$3	$4	$10	$12
Percent Decrease				

If there were 730 students taking a mathematics class last year, and there are 892 taking a mathematics class this year, what is the percent change in the number of students taking a mathematics class, rounded to the nearest tenth of a percent?

There were 730 students initially and number of students increased to 892, so this is a percent increase problem.

The amount of increase is $892 - 730$, or 162. Use the percent increase equation to find the percent increase.

$$\text{Amount of Increase} = \text{Percent Increase} \cdot \text{Original Value}$$

$$162 = x \cdot 730$$
$$\frac{162}{730} = \frac{730x}{730}$$
$$0.222 \approx x$$

The percent increase was 22.2%, rounded to the nearest tenth of a percent.

Determine the percent change. (Round to the nearest tenth of a percent.)
3. Last week a steak cost $19.50. This week a steak costs $23.45.

4. Twenty years ago, a movie ticket cost $3. This year is costs $8.

5. In 2018 the population of Valdosta, GA was 56,085 people. In 1995 it was 43,005 people.

Comparing Comparisons: Difference, Ratio and Relative Difference

Two quantities which are measured in the same unit can be compared by answering the following questions:
- The first is how much larger than the second?
- The first is how many times as large as the second?
- The first is how many times larger than the second?

The question, "how much larger," is answered by finding the difference,

$$\text{first} - \text{second}$$

The question, "how many times as large as," is answered by computing the ratio,

$$\frac{\text{first}}{\text{second}}$$

The answer to the question of "how many times larger" is the relative difference,

$$\frac{\text{first} - \text{second}}{\text{second}}$$

Differences are measured in the same unit in which the two quantities are measured. The ratios and relative differences are real numbers with no units, and are often expressed as percents.

A group of 36 workers was asked whether they work in the private or public sector. The results are displayed in the following bar graph.

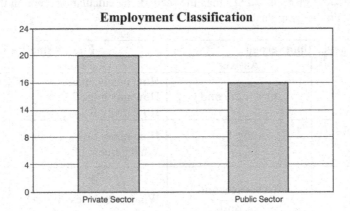

1. The number of workers who work in the private sector is how much larger than the number who work in the public sector? Show a calculation, answer in a complete sentence and include the appropriate unit.

2. The number of workers who work in the private sector is how many times as large as the number who work in the public sector? Express as a mixed number and as a decimal, then complete the sentence below.

The number who work in the private sector is _____% as large as the number who work in the public sector. Confirm by comparing the heights of the bars in the graph above.

The bar graph below includes a bar that measures the difference between the numbers of workers who work in the private and the public sectors.

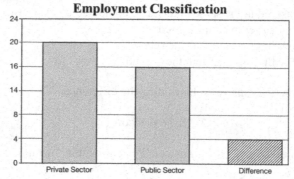

Employment Classification

3. The difference between the numbers of workers who work in the private and public sectors is how many times as large as the number who work in the public sector? Express as a simplified fraction and as a decimal, then complete the sentence below.

   The number who work in the private sector is _____% larger than the number who work in the public sector. Confirm by comparing the heights of the bar for the difference to the height of the bar for public sector.

In the three comparisons above, the first quantity mentioned in the comparison (the number who work in the private sector) was larger than the second quantity (the number who work in the public sector). When the first quantity mentioned in a comparison is *smaller* than the second, the calculations remain the same, but the wording is altered accordingly, as in the table below.

First is larger than second		First is smaller than second	
Question	Answer	Question	Answer
How much larger? How much more? How many more?	first − second	How much smaller? How much less? How many fewer?	OPPOSITE of: first − second
How many times as large? How many times the size? What percent of?	$\dfrac{\text{first}}{\text{second}}$	How many times the size? What percent of?	$\dfrac{\text{first}}{\text{second}}$
How many times larger? What percent larger?	$\dfrac{\text{first − second}}{\text{second}}$	What percent less/fewer?	OPPOSITE of: $\dfrac{\text{first − second}}{\text{second}}$

4. The number of workers who work in the public sector is how much smaller than the number who work in the private sector? Show a calculation, answer in a complete sentence and include the appropriate unit.

5. The number of workers in the public sector is how many times the size of the number who work in the private sector? Express as a simplified fraction and as a decimal, then complete the sentence below.

   The number who work in the public sector is _____% the size of the number who work in the private sector. Confirm by comparing the heights of the bars in the graph above.

6. The difference between the number who work in the public sector and the private sector is how many times as large as the number who work in the private sector? Express as a simplified fraction and as a decimal, then complete the sentence below.

   The number who work in the public sector is _____% smaller than the number who work in the private sector. Confirm by comparing the heights of the bars in the graph above.

Of the sixteen workers who work in the public sector, twelve of them drive alone to work in their own vehicles. The other four take public transportation or carpool to work.

7. Make a bar graph with bars for public transit & carpool, private vehicle, and a third bar for the difference between the two. Title the graph.

8. Show the relevant computation, then fill in each blank with a number. Confirm your answer by comparing the heights of the bars in the bar graph for this data.

   a. _____ more public sector workers drive themselves than take public transportation or carpool.

   b. _____ times as many drive themselves than take public transportation or carpool.

   c. The number who drive themselves is _____% larger than the number who take public transportation or carpool.

   d. _____ fewer public sector workers take public transportation or carpool than drive themselves.

   e. _____% as many take public transportation or carpool than drive themselves.

   f. The number who take public transportation or carpool is _____% smaller than the number who drive themselves.

The bar graph below shows the transportation choices of the twenty private sector workers.

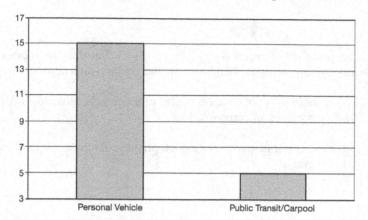

Private Sector Workers' Mode of Transportation

9. How many more take their personal vehicle to work than use public transportation or carpool? Show a calculation, answer in a complete sentence and include the appropriate unit.

       Confirm: Is the difference between the heights of the bars in the graph the same as the difference between the numbers of workers those heights represent?

10. The number of workers who take their personal vehicle to work is how many times as large as the number who use public transportation or carpool? Show a calculation. Use the answer to fill in the blank in the question below.

       Confirm: Is the bar that represents the workers who drive their own vehicle _____ times as tall as the bar that represents the workers who take public transportation or carpool? If not, how many times taller is the bar for personal vehicle than the bar for public transportation or carpool?

This example illustrates the fact that for ratios to make sense, there must be a true zero. The vertical axis of the bar graph must start at zero for the ratios of the heights of the bars to be equal to the ratios of the numbers their heights represent.

11. Draw a new bar graph, for the same data, in which the ratios of the heights of the bars are equal to the ratios of the numbers those heights represent.

Proportional Proportions Puzzles

If a sample of individuals from a population is surveyed and the sample is representative of the entire population, the number of people in the whole population who would have chosen a specific response, if asked, is approximately proportional to the number of those surveyed who gave that response.

1. A sample of 120 students who attend a college with a total enrollment of 14,500 was surveyed. The students who were surveyed were asked to pick their favorite genre of music from the list of categories shown in the table. Students could choose only one genre. Assuming that the sample is representative of the student body, complete the table with the best approximate values.

Favorite Music Genre	Percent	Students who were Surveyed	Entire Student Body
Country	20		
Jazz			
Pop		30	
Rhythm & Blues			4350
Other	15		
Total:	100	120	14,500

2. Six hundred residents from a small town with 22,050 registered voters were polled to see how they felt about the idea of a new stoplight being installed at a busy intersection on the main highway through town. Complete the table, assuming that the sample of residents surveyed was representative of the population of all registered voters in the town.

	Percent	Residents who were Surveyed	All Registered Voters
Strongly in favor	8		
Favor, but with reservations			5733
Ambivalent		96	
Oppose, but if the conditions are modified, could be persuaded			
Strongly oppose	18		
Total:			

3. Twenty-five customers who visited a fast-food restaurant on the same day were surveyed. They were asked to say "Yes" or "No" to indicate which of the descriptions applied to them. The table below looks very similar to those above. What about this table makes it impossible to complete?

	Percent	Customers who were Surveyed	All Customers that Day
Repeat Customer	8		
Satisfied with the Service			882
Satisfied with Food Quality		16	
Satisfied with Wait Time			
Would Return Again	12		
Total:			1260

4. Some samples of individuals to be surveyed represent their populations better than others. If the manager of a fast food restaurant wanted to know about all customers' experiences at the restaurant on a particular day, describe ways in which she could select 25 customers to survey in a way that would...
    a. accurately represent all the customers who visited that day.

    b. not accurately represent all the customers who visited that day.

5. Of the 25 customers who were surveyed at the fast food restaurant, 15 of them ordered coffee.
    a. About how many of the day's 1260 customers would you expect to have ordered coffee?

    b. Only 235 of the day's customers actually did order coffee. Why might the estimate from part a. have been so high?

Measurement

In the United States (US) we tend to use feet, pounds, and gallons to measure the length, weight, and capacity of an object. However, there are many other related units in case one of these are not a good size. There are inches, yards, and miles if feet are not practical; ounces, stones, and tons if pounds are not practical; and teaspoons, tablespoons, cups and quarts if gallons are not practical.

Length: 12 inches = 1 foot, 3 feet = 1 yard, 5280 feet = 1 mile

Weight: 16 ounces = 1 pound, 14 pounds = 1 stone, 2000 pounds = 1 ton

Capacity (or Volume):   3 teaspoons = 1 tablespoon, 16 tablespoons = 1 cup, 4 cups = 1 quart, 4 quarts = 1 gallon, 16 cups = 1 gallon

Suppose you wanted to convert 42 inches to feet, to yards, and to miles.

To convert to feet, multiply $42 \text{ inches} \cdot \dfrac{1 \text{ foot}}{12 \text{ inches}} = 3.5 \text{ feet}$.

To convert to yards, multiply $3.5 \text{ feet} \cdot \dfrac{1 \text{ yard}}{3 \text{ feet}} \approx 1.167 \text{ yards}$.

To convert to miles, multiply $3.5 \text{ feet} \cdot \dfrac{1 \text{ mile}}{5280 \text{ feet}} \approx 0.00066 \text{ miles}$.

1. Convert each quantity to the desired unit. Round to the nearest hundredth, if necessary.

	inches	feet	yards
a. 2.1 miles			

	ounces	stones	tons
b. 1673 pounds			

	teaspoons	tablespoons	gallons
c. 7 cups			

Suppose we are increasing a recipe for a party. The original recipe calls for five teaspoons of cocoa powder. If we are making a batch 12 times as large, how much cocoa do we need? One answer is $5 \cdot 12 = 60$ teaspoons, but that will be tedious to measure out. That is equivalent to 20 tablespoons, which would still be tedious to measure out. However, 20 tablespoons is equivalent to 1 cup and 4 tablespoons, which would be easier to measure out.

2. Convert each of the following measurements to the indicated units.

a. 62 inches to feet and inches   b. 129 ounces to pounds and ounces

c. 122 teaspoons to cups and teaspoons   d. 560 cups to gallons and cups

Most of the world uses the metric system for their measurements. In the metric system, the basic unit of length is the meter; the basic unit of mass is the gram; and the basic unit of capacity is the liter. One appeal of the metric system is that it is based on powers of 10.

Length: 10 millimeters = 1 centimeter, 100 centimeters = 1 meter, 1000 meters = 1 kilometer
Mass: 1000 milligrams = 1 gram, 1000 grams = 1 kilogram
Capacity (or Volume): 1000 milliliters = 1 liter

3. Convert the given quantity to the desired unit.

	meters	kilometers	millimeters
a. 42 centimeters			

	milligrams	kilograms
b. 1673 grams		

Sometimes we need to convert between the US system and the metric system.
Conversion Factors Between the U.S. and Metric Systems
Length

English to Metric	*Metric to English*
1 mile ≈ 1.61 kilometers	1 kilometer ≈ 0.62 miles
1 foot ≈ 0.305 meters	1 meter ≈ 3.28 feet
1 inch = 2.54 centimeters	1 centimeter ≈ 0.39 inches

Weight/Mass

English to Metric	*Metric to English*
1 gallon ≈ 3.785 liters	1 liter ≈ 0.264 gallons

Capacity

English to Metric	*Metric to English*
1 pound ≈ 0.454 kilograms	1 kilogram ≈ 2.2 pounds
1 ounce ≈ 28.35 grams	1 gram ≈ 0.035 ounces

4. Compute the following conversions:
a. 17 liters to gallons

b. 383 kilograms to pounds

c. 4 miles to kilometers

d. 117 inches to meters

5. Which would be longer, 55 meters or 55 yards? Explain your answer.

6. A patient weighing 156 pounds is to receive 0.05 milligram per kilogram of a drug. If each tablet contains 0.5 milligrams, then how many tablets should the patient receive?

7. A children's medicine says to give 10 milliliters per kilogram. If the child weighs 48 pounds, how many milliliters should the child get?

Real Numbers

Any real number that can be expressed as a fraction whose numerator and denominator are integers is a rational number. A square root of a nonnegative number that can be simplified to be an integer or a fraction, such as

$\sqrt{25} = 5$ or $\sqrt{\dfrac{49}{4}} = \dfrac{7}{2}$, is a rational number.

Otherwise the square root is an irrational number. For example, $\sqrt{2}$ is an irrational number.

1. Classify the following numbers as rational or irrational.

a. $\dfrac{3}{5}$　　　　　b. 3.8　　　　　c. $\sqrt{7}$

d. $\sqrt{49}$　　　　　e. $\dfrac{\sqrt{5}}{4}$　　　　　f. $\sqrt{\dfrac{48}{27}}$

To estimate a square root without using the square root function on a calculator, begin by determining the two consecutive integers that the square root lies between. For example, since 20 is between the perfect squares 16 and 25, $\sqrt{16} < \sqrt{20} < \sqrt{25}$. Therefore, $4 < \sqrt{20} < 5$.

Square each integer, and find the distance between the two squares. The distance between 16 and 25 is 9.

So, 20 is $\dfrac{4}{9}$ of the way between 16 and 25.

Since $\dfrac{4}{9} \approx 0.44$, $\sqrt{20}$ can be estimated to be 0.44 larger than $\sqrt{16}$. In other words, an estimate of $\sqrt{20}$ is 4.44.

Using the built-in function for approximating square roots, $\sqrt{20} \approx 4.47$, so the estimate of 4.44 is a good estimate.

2. Estimate each square root to the nearest hundredth using the above procedure. Then approximate the square root to the nearest hundredth using the built-in function on a calculator. Determine how far off your estimate was from the approximated value.

Square Root	Estimate	Approximation (Calculator)	Difference
a. $\sqrt{13}$			
b. $\sqrt{32}$			
c. $\sqrt{85}$			
d. $\sqrt{200}$			

3. Insert a symbol, $<$ or $>$, between the two numbers given to make a true inequality. Do not use a calculator. Explain your reasoning.

a. $\sqrt{50}$ 7

b. $\sqrt{17}$ 4

c. 3 $\sqrt{5}$

4. Use a calculator to evaluate the following expressions that involve real numbers. As an extra challenge, do so by pressing "Enter" only once. Round non-integer results to two decimal places.

a. $\sqrt{(5-3)^2 + (1.5-4)^2}$

b. $1000\left(1 + \frac{0.06}{12}\right)^{12 \times 1.25}$

c. $\frac{15}{3(15-3)}$

d. $\frac{(6-2)^2 + (2.1-2)^2 + (4-2)^2}{3}$

e. $\sqrt{\frac{0.12 \times 0.88}{24}}$

f. $\sqrt{\frac{(8-3)^2}{5}}$

Algebraic Expressions

To evaluate an algebraic expression for a given value, replace the variable with the value and then simplify.

Evaluate the expression $6x^2 - 3x + 7$ when …

$x = 2$	$x = -1$
$6x^2 - 3x + 7$	$6x^2 - 3x + 7$
$6(2)^2 - 3(2) + 7$	$6(-1)^2 - 3(-1) + 7$
$= 6(4) - 3(2) + 7$	$= 6(1) - 3(-1) + 7$
$= 24 - 6 + 7$	$= 6 + 3 + 7$
$= 25$	$= 16$

1. Evaluate the expression $3x^2 - 7x + 1$ for the following values:

a. $x = 1$ b. $x = 2$ c. $x = -3$ d. $x = -4$

2. Evaluate each of the following expressions for $x = 3$:

a. $2x^3 + 3x^3 - 1$ b. $4x^2 + 5x - x^2 - 2x - 6$ c. $5x^3 - 1$ d. $3x^2 + 3x - 6$

Notice the pairs in exercise 2 that gave the same answer. This is because the pairs are equivalent when you combine like terms.

$$2x^3 + 3x^3 - 1$$
$$= \underbrace{2x^3 + 3x^3}_{2+3=5} - 1$$
$$= 5x^3 - 1$$

$$4x^2 + 5x - x^2 - 2x - 6$$
$$= 4x^2 - x^2 + 5x - 2x - 6$$
$$= \underbrace{4x^2 - x^2}_{4-1=3} \underbrace{+5x - 2x}_{5-2=3} - 6$$
$$= 3x^2 + 3x - 6$$

Like terms are terms whose variable(s) and exponent(s) are the same.
- $3x^4$ and $7x^4$ are like terms – same variable and same exponent
- $3x^4$ and $7x^3$ are NOT like terms – same variable but different exponents
- $3x^2y^3$ and $-4x^2y^3$ are like terms – same variables with the same exponents
- $3x^3y^2$ and $4x^2y^3$ are NOT like terms – same variables but they have different exponents (even though they are both degree five)

3. Simplify each expression by combining like terms.

a. $5x - 2x - 6x$

b. $3x + 2x - 5y + 3y$

c. $3x^3 - y^3 + x^3 - 2y^3$

d. $3x^2 + 3x^3$

e. $3\dfrac{x^3}{r} - \dfrac{y^3}{r} + \dfrac{y^3x^3}{r} - \dfrac{2x^3}{r} + \dfrac{y^3}{r}$

4.

a. List two like terms that can be combined to be $7x^2$.

b. List three like terms that can be combined to be $-19x$.

c. List four terms that can be combined to be $-9x + y$.

Solving Linear Equations

To understand how to solve a linear equation, it is helpful to realize how it is associated to evaluating a linear algebraic expression.

Evaluate $7x - 4$ when $x = 5$.		Solve $7x - 4 = 31$	
$7x - 4$			
$7(5) - 4$	Substitute 5 for x.	$7x - 4 = 31$	
$= 35 - 4$	Multiply 5 by 7.	$7x = 35$	Add 4 to both sides of the equation.
$= 31$	Subtract 4 from 31.	$x = 5$	Divide both sides of the equation by 7.

Notice that the steps used to solve the equation are the inverse operations of those used to simplify the expression, and they are performed in the opposite order.

Instead of multiplying by 7 and then subtracting 4 when evaluating the expression, we first add 4 and then divide by 7.

This strategy can be helpful when deciding which step to take when solving a linear equation.

This strategy is similar to reversing directions. If you drove 10 miles north, turned right, then drove 5 miles east to get to school, then if you reverse those directions you would return to your starting point.

First reverse or undo the last part "drove 5 miles east" by driving 5 miles west.

Then undo the right turn by turning left.

Finally undo the "drove 10 miles north" by driving 10 miles south.

Finish

Start

1.

a. List the two steps to evaluate $9x + 2$ when $x = -4$.

(Only list the steps. Do not actually evaluate.)

1.

2.

b. Using part a, list the two steps that can be used to solve $9x + 2 = -34$.

(Only list the steps. Do not solve.)

1.

2.

c. Use the steps from part b to solve: $9x + 2 = -34$

2.

a. List the three steps to evaluate $2(5x+11)$ when $x=8$. (Only list the steps. Do not actually evaluate.)	b. Using part a, list the three steps that can be used to solve $2(5x+11)=102$. (Only list the steps. Do not solve.)
1.	1.
2.	2.
3.	3.

c. Use the steps from part b to solve: $2(5x+11)=102$

3.

a. List the two steps to evaluate $\left(\dfrac{5}{9}\right)\cdot(F-32)$ when $F=50$. (Only list the steps. Do not actually evaluate.)	b. Using part a, list the two steps that can be used to solve $\left(\dfrac{5}{9}\right)\cdot(F-32)=10$. (Only list the steps. Do not solve.)
1.	1.
2.	2.

c. Use the steps from part b to solve: $\left(\dfrac{5}{9}\right)\cdot(F-32)=10$

4. Write an equation that can be solved by adding 5 to both sides of the equation and then dividing both sides of the equation by 2.

5. Write an equation that can be solved by subtracting 7 from both sides of the equation and then multiplying both sides of the equation by 3.

6. Write an equation that can be solved by dividing both sides of the equation by 4, then adding 13 to both sides of the equation, and then dividing both sides of the equation by 5.

7. Write an equation whose solution is $x = 9$ that can be solved by adding 3 to both sides of the equation and then dividing both sides of the equation by 5.

8. Write an equation whose solution is $x = -6$ that can be solved by subtracting 1 from both sides of the equation and then dividing both sides of the equation by 8.

9. Write an equation whose solution is $x = 4$ that can be solved by dividing both sides of the equation by 2, then subtracting 11 from both sides of the equation, and then dividing both sides of the equation by 3.

Linear Inequalities

The statement "3 is less than 5" $(3 < 5)$ is equivalent to the statement "5 is greater than 3" $(5 > 3)$.

This property can be used to graph the solutions of an inequality like $-2 < x$. Rather than thinking about the inequality as -2 is less than x, it will be easier to think about it as x is greater than -2. This inequality can be rewritten as $x > -2$, and the solutions can be displayed on the number line by shading to the right of an open circle at -2.

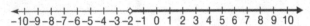

1. Rewrite the inequality so the variable x is on the left side of the inequality, and graph the solutions.
a. $5 > x$ b. $7 \leq x$ c. $-6 \geq x$

The same property can be used to help understand why we change the direction of an inequality sign when multiplying or dividing both sides of the inequality by a negative number.

Suppose you wanted to solve the inequality $-x > 6$. One approach is the following.

$-x > 6$
$\quad 0 > x + 6$ *Add x to both sides of the inequality, so the coefficient of the variable term is positive.*
$\quad -6 > x$ *Subtract 6 from both sides of the inequality to isolate the variable term on the right side of the inequality.*
$\quad x < -6$ *Rewrite the inequality.*

Notice that each side of the solution, $x < -6$, can be obtained from the original inequality by dividing both sides of the inequality $-x > 6$ by -1. Also notice that the direction of the inequality sign has changed from $>$ to $<$. So, when multiplying or dividing both sides of an inequality by a negative number, you must change the direction of the inequality sign.

2. Solve the inequality.

a. $-2x < -20$

b. $\dfrac{x}{-4} \geq 3$

3. Solve the inequality.
a. $-6x + 25 < -11$

b. $-3x - 17 \geq 19$

Another important concept involving inequalities is determining whether an endpoint is included with the solutions of an inequality or not.

4. Is the number 8 included as a solution of the given inequality? Explain why or why not.
a. $x < 8$

b. $x \leq 8$

c. $x \geq 8$

d. $x > 8$

5. Is the number 12 included as a solution of the given inequality? Explain why or why not.
a. x is at least 12

b. x is more than 12

c. x is less than 12

d. x is 12 or fewer

6. Explain how to determine whether to use open or closed circles when graphing the solutions of an inequality on a number line.

7. Explain how to determine whether to use parentheses or square brackets when expressing the solutions of an inequality using interval notation.

Composite Figures

The area of a two-dimensional figure is a measure of the amount of space contained inside the figure.

Area Formulas

Rectangle with length L and width W:	$A = L \cdot W$
Square with side x:	$A = x^2$
Triangle with base b and height h:	$A = \frac{1}{2}bh$
Circle with radius r:	$A = \pi r^2$

A **composite figure** is a figure that can be made of two or more basic figures. The area of a composite figure can be found by finding the area of each basic figure and then finding the total of those areas.

The composite figure on the left is comprised of two rectangles, as shown in the figure on the right.

Denote the area of the left rectangle as A_1 and the area of the rectangle on the right as A_2.

The length of the rectangle on the left is 5 cm and the width is 4 cm.

$$A_1 = L \cdot W$$

$$A_1 = 5 \cdot 4 = 20 \text{ cm}^2$$

The length of the rectangle on the right is 3 cm, which can be found by subtracting $8 \text{ cm} - 5 \text{ cm}$.

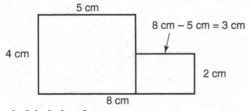

The width of the rectangle is already labeled as 2 cm.

$$A_2 = L \cdot W$$

$$A_2 = 3 \cdot 2 = 6 \text{ cm}^2$$

The area of the composite figure can be found by adding $A_1 + A_2$, which is 26 cm^2.

By the way, the composite figure could also be divided into two different rectangles as shown.

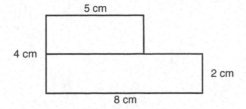

Use the appropriate formula(s) to find the area of the composite figures. Use $\pi \approx 3.14$ and round to the nearest hundredth.

1.

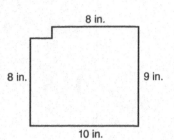

2.

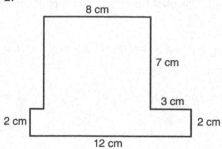

3.

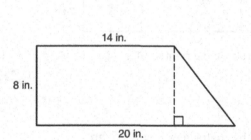

4.

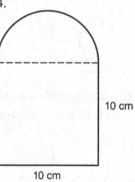

5.

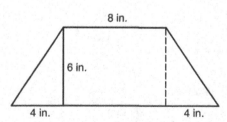

6.

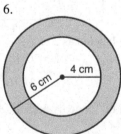

7. A **regular pentagon** is a 5-sided figure whose sides all have the same length. Find the area of the pentagon, in terms of x and h, by dividing it into five triangles.

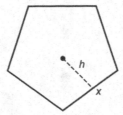

(Pentagon, 5 equal sides)

8. A **regular hexagon** is a 6-sided figure whose sides all have the same length. Find the area of the hexagon, in terms of x and h, by dividing it into six triangles.

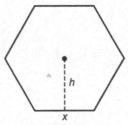

(Hexagon, 6 equal sides)

9. A **regular octagon** is an 8-sided figure whose sides all have the same length. Find the area of the octagon, in terms of x and h, by dividing it into eight triangles.

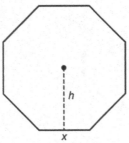

(Octagon, 8 equal sides)

10. Based on your results from exercises 7 through 9, determine a formula for the formula of a **regular n-gon** (n-sided figure with n equal sides) with sides of length x and distance h from the center of the n-gon to the center of each side.

Plotting Points

A crucial skill when plotting ordered pairs is being able to determine how to label the x- and y-axes. Determine the minimum and maximum values for x and y. Label the axes so that each ordered pair fits on the graph. Look for patterns in the data. For example, if all of the x-coordinates are multiples of 5 consider using a scale of 5.

1. Consider the ordered pairs $(-18,4)$, $(6,16)$, $(0,-12)$, $(-10,2)$, and $(14,8)$.

a. What are the minimum and maximum values of x?

b. What are the minimum and maximum values of y?

c. Is there a pattern in the x- and y-coordinates? Rather than using a scale of 1, what would be a good scale to use?

d. Create a rectangular coordinate plane with the scale determined in part c. Plot the ordered pairs on the plane.

2. Create a rectangular coordinate plane and plot the ordered pairs: $(-45,-45)$, $(50,20)$, $(25,-30)$, $(-10,-35)$, $(0,5)$, $(-30,0)$, $(40,50)$, and $(10,-10)$.

3. Create a rectangular coordinate plane and plot the ordered pairs: $(700,400)$, $(900,-100)$, $(200,-600)$, $(-100,300)$, $(-400,-300)$, $(200,1000)$, $(-300,-800)$.

4. Complete the table for the equation $y = 3x - 8$, plot the ordered pairs, and use them to graph the line.

x	$y = 3x - 8$
-2	
0	
2	
4	
6	
8	

5. Complete the table for the equation $y = -5x - 20$, plot the ordered pairs, and use them to graph the line.

x	$y = -5x - 20$
-10	
-5	
0	
5	

6. Complete the table for the equation $y = 4x + 12$, plot the ordered pairs, and use them to graph the line.

x	$y = 4x + 12$
-9	
-6	
-3	
0	
3	

Intercepts

The **x-intercept** of a line is the point where the graph of the line crosses the *x*-axis.

1. Plot a point on the *x*-axis and label its coordinates as an ordered pair. Draw a line that passes through that point. The point is the *x*-intercept of the line. Repeat this for two other points on the *x*-axis, creating the graphs of a total of three lines.

2. Other than being *x*-intercepts, what do the three points in exercise 1 have in common?

The *y*-coordinate of any *x*-intercept is always equal to 0. To find the *x*-intercept of any line, substitute 0 for *y* and solve for *x*.

3. Find the *x*-intercept of the line. Do not graph the line.
a. $3x + 4y = -36$ b. $-7x + 6y = -126$ c. $4x - 3y = 18$

The **y-intercept** of a line is the point where the graph of the line crosses the *y*-axis.

4. Plot a point on the *y*-axis and label its coordinates as an ordered pair. Draw a line that passes through that point. The point is the *y*-intercept of the line. Repeat this for two other points on the *y*-axis, creating the graphs of a total of three lines.

Notice that the three points in exercise 4 all have a *x*-coordinate of 0. The *x*-coordinate of any *y*-intercept is always equal to 0. To find the *y*-intercept of any line, substitute 0 for *x* and solve for *y*.

5. Find the *x*-intercept of the line. Do not graph the line.
a. $3x + 4y = -36$ b. $-7x + 6y = -126$ c. $4x - 3y = 18$

6. Draw a graph that has an x-intercept, but does not have a y-intercept.

7. What type of line is the line you drew in exercise 6?

8. Draw the graph of a line that has an x-intercept at $(6,0)$ but does not have a y-intercept.

9. List the coordinates of three other points on the line you graphed in exercise 8. What do the points have in common?

Since all of the points listed in exercise 9 have x-coordinates of 6, the equation of the line is $x=6$. (x is 6 for every point on the line.) In general, the equation of a vertical line with an x-intercept of $(a,0)$ is $x=a$.

10. What type of line has a y-intercept, but does not have an x-intercept.

11. Draw the graph of a line that has a y-intercept at $(0,4)$ but does not have an x-intercept.

12. List the coordinates of three other points on the line you graphed in exercise 11. What do the points have in common?

Since all of the points listed in exercise 12 have y-coordinates of 4, the equation of the line is $y=4$. (y is 4 for every point on the line.) In general, the equation of a horizontal line with a y-intercept of $(0,b)$ is $y=b$.

13. Is it possible to graph a line whose x-intercept and y-intercept are the same point? If so, draw an example of a line with this property.

14. Draw two other lines whose x-intercept and y-intercept are the same point.

15. What point do the graphs of the three lines in exercises 13 and 14 have in common?

16. Draw the graph of a line that passes through the origin, $(0,0)$, as well as the point $(1,2)$.

17. Notice that the line you graphed in exercise 16 also passes through the points $(2,4)$, $(3,6)$, $(5,10)$, $(-1,-2)$, and $(-4,-8)$. What is the relation between the y-coordinates and x-coordinates of these points?

Analyzing Slope: Sliding Into More Facts

1. Complete the ordered pairs for P1 and P2, then apply the slope formula accordingly.

A.

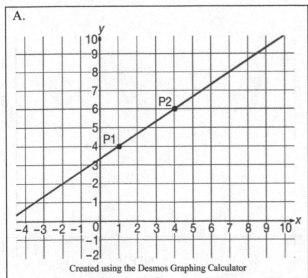

Created using the Desmos Graphing Calculator

P1: (,   ) and P2: (,   )

$m =$

B.

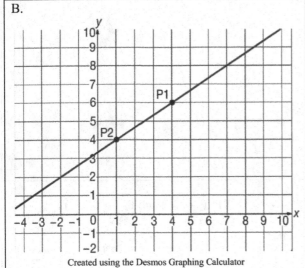

Created using the Desmos Graphing Calculator

P1: (,   ) and P2: (,   )

$m =$

C.

Created using the Desmos Graphing Calculator

P1: (,   ) and P2: (,   )

$m =$

D.

Created using the Desmos Graphing Calculator

P1: (,   ) and P2: (,   )

$m =$

E.

Created using the Desmos Graphing Calculator

P1: (,   ) and P2: (,   )

$m =$

2.   Name a pair of graphs A through E, that, when considered together, illustrate each fact.

Fact	Graphs
When using the slope formula, the result will be the same regardless of which of the same two points is taken as P1 and which is taken as P2.	
The slopes of non-vertical parallel lines are equal.	
The slope formula, when applied to different pairs of points on the same non-vertical line, gives the same result.	

3.   Confirm that the two lines in Graph E are perpendicular.

Multiply the slope of the solid line by the slope of the dashed line in Graph E.

Does the result support the statement that follows?

If two non-vertical lines are perpendicular, the product of their slopes is – 1.

4.   Graph the line that passes through the point (5,3) with slope $m = -\frac{2}{3}$, as instructed.

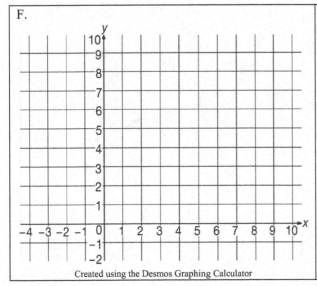

F.

Plot the point (5,3) and label it as P1. From that point, plot and label the point P2 obtained by applying $m = \frac{2}{-3}$

Give the coordinates of P2: (,   )

From the point P1: (5,3), plot and label the point P3 obtained by applying
$m = \frac{-2}{3}$

Give the coordinates of P3: (,   )

Do the three points form a line? If so, draw the line that passes through them.

Created using the Desmos Graphing Calculator

Does Graph F illustrate the statement that follows?

For all real numbers a and b with $b \neq 0$,

$$-\frac{a}{b} = \frac{a}{-b} = \frac{-a}{b}$$

Equations of Lines

We will explore several ways to find the equation of a line. We will be using the following line.

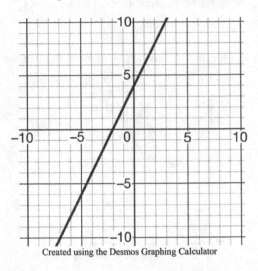

Created using the Desmos Graphing Calculator

Strategy 1 – Slope and *y*-intercept are known

If the slope m and *y*-intercept $(0,b)$ of a line are known, find the equation of the line by substituting values for m and b in the slope-intercept form of a line: $y = mx + b$.

The line has its *y*-intercept at $(0,4)$, so $b = 4$.

The slope can be determined by measuring the rise and run from one point on the line to another. Notice that the point $(1,6)$ is also on the line, and that is 2 units above and 1 unit to the right of the *y*-intercept. The slope is $m = \dfrac{2}{1}$, or $m = 2$.

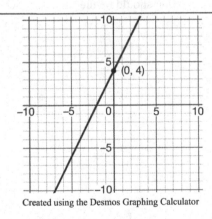

Created using the Desmos Graphing Calculator

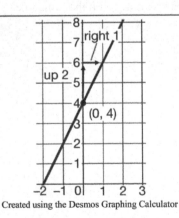

Created using the Desmos Graphing Calculator

1. Find the equation of a line whose *y*-intercept is $(0,4)$ and slope is $m = 2$.

<u>Strategy 2 – Slope and a point on the line are known</u>

If the slope m of a line, m, and a point $\left(x_1, y_1\right)$ on the line are known, find the equation of the line by substituting values for m, x_1, and y_1 in the point-slope form of a line: $y - y_1 = m\left(x - x_1\right)$. After substituting, rewrite the equation in slope-intercept form by solving for y.

The line we are working with has a slope $m = 2$, and it passes through the point $(1, 6)$.

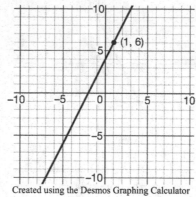

Created using the Desmos Graphing Calculator

2. Find the equation of a line whose slope is $m = 2$ and passes through the point $(1, 6)$.

Since we were working with the same line in exercises 1 and 2, the equation should be the same for each exercise.

<u>Strategy 3 – Two points on the line are known</u>

If two points (x_1, y_1) and (x_2, y_2) on the line are known, find the equation of the line by first computing the slope of the line using the formula $m = \dfrac{y_2 - y_1}{x_2 - x_1}$. Then substitute values for m, x_1, and y_1 in the point-slope form of a line: $y - y_1 = m(x - x_1)$. (You can use the coordinates of either point when substituting into the point-slope form.). After substituting, rewrite the equation in slope-intercept form by solving for y.

The line we are working with passes through the points $(1,6)$ and $(2,8)$.

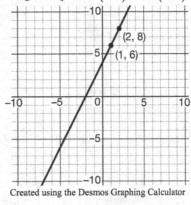

Created using the Desmos Graphing Calculator

3. Find the slope of the line that passes through the points $(1,6)$ and $(2,8)$.

4. Find the equation of the line that passes through the points $(1,6)$ and $(2,8)$.

Again, since we were working with the same line as we used in exercises 1 and 2, the equation should be the same for each exercise.

For exercises 5-7, use the following graph.

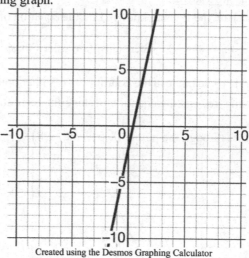

Created using the Desmos Graphing Calculator

5.
a. Find the *y*-intercept of the line from the graph.   b. Determine the slope of the line from its graph.

c. Use the results from parts a and b and the slope-intercept form of a line, $y = mx + b$, to determine the equation of the line.

6.
a. Find the coordinates of a point on the line (other than the *y*-intercept).

b. Use the slope from exercise 5 and the point from part a to determine the equation of the line. Write the equation in slope-intercept form.

7.
a. Find the coordinates of two points on the line.   b. Compute the slope of the line using the slope formula.

c. Use the slope and one of the points from part a to determine the equation of the line. Write the equation in slope-intercept form.

Rules for Exponents

Although having the exponent rules memorized, being able to derive the rules from examples is an excellent strategy for when the rule escapes your memory.

<u>Product Rule</u>

Simplify $x^5 \cdot x^3$.

$$x^5 \cdot x^3$$
$$= (x \cdot x \cdot x \cdot x \cdot x) \cdot (x \cdot x \cdot x)$$
$$= x^8$$

So, $x^5 \cdot x^3 = x^{5+3} = x^8$.

This can be used to recall the product rule:
For any base x, $x^m \cdot x^n = x^{m+n}$.

1. Create an example of your own, and use it to show that for any base x, $x^m \cdot x^n = x^{m+n}$.

<u>Power Rule</u>

Simplify $\left(x^5\right)^3$.

$$\left(x^5\right)^3$$
$$= x^5 \cdot x^5 \cdot x^5$$
$$= x^{5+5+5}$$
$$= x^{15}$$

So, $\left(x^5\right)^3 = x^{5 \cdot 3} = x^{15}$.

This can be used to recall the power rule:
For any base x, $\left(x^m\right)^n = x^{m \cdot n}$.

2. Create an example of your own, and use it to show that for any base x, $\left(x^m\right)^n = x^{m \cdot n}$.

<u>Power of a Product Rule</u>

Simplify $(xy)^3$.

$$(xy)^3$$
$$= xy \cdot xy \cdot xy$$
$$= x \cdot x \cdot x \cdot y \cdot y \cdot y$$
$$= x^3 y^3$$

So, $(xy)^3 = x^3 y^3$.

This can be used to recall the power of a product rule:
For any base x, $\left(x^m\right)^n = x^{m \cdot n}$.

3. Create an example of your own, and use it to show that for any bases x and y, $(xy)^n = x^n y^n$.

Quotient Rule

Simplify $\dfrac{x^5}{x^3}$.

$$\frac{x^5}{x^3} = \frac{x \cdot x \cdot x \cdot x \cdot x}{x \cdot x \cdot x}$$

$$= \frac{\overset{1}{\cancel{x}} \cdot \overset{1}{\cancel{x}} \cdot \overset{1}{\cancel{x}} \cdot x \cdot x}{\underset{1}{\cancel{x}} \cdot \underset{1}{\cancel{x}} \cdot \underset{1}{\cancel{x}}}$$

$$= x \cdot x$$

So, $\dfrac{x^5}{x^3} = x^{5-3} = x^2$.

This can be used to recall the quotient rule:

For any nonzero base x, $\dfrac{x^m}{x^n} = x^{m-n}$.

4. Create an example of your own, and use it to show that for any nonzero base x, $\dfrac{x^m}{x^n} = x^{m-n}$.

Zero Exponent Rule

Simplify $\dfrac{x^5}{x^5}$.

$$\frac{x^5}{x^5} = x^{5-5} = x^0$$

Also, any nonzero number divided by itself is equal to 1.

$$\frac{x^5}{x^5} = 1$$

Since $\dfrac{x^5}{x^5} = x^0$ and $\dfrac{x^5}{x^5} = 1$, thus $x^0 = 1$.

This can be used to recall the zero exponent rule:

For any nonzero base x, $x^0 = 1$.

5. Create an example of your own, and use it to show that for any nonzero base x, $x^0 = 1$.

Power of a Quotient Rule

Simplify $\left(\dfrac{x}{y}\right)^3$.

$$\left(\frac{x}{y}\right)^3 = \frac{x}{y} \cdot \frac{x}{y} \cdot \frac{x}{y} = \frac{x \cdot x \cdot x}{y \cdot y \cdot y} = \frac{x^3}{y^3}$$

So, $\left(\dfrac{x}{y}\right)^3 = \dfrac{x^3}{y^3}$.

This can be used to recall the power of a quotient rule:

For any bases x and y, $\left(\dfrac{x}{y}\right)^n = \dfrac{x^n}{y^n}$ $(y \neq 0)$.

6. Create an example of your own, and use it to show that for any bases x and y, $\left(\dfrac{x}{y}\right)^n = \dfrac{x^n}{y^n}$ $(y \neq 0)$.

Negative Exponent Rule

Simplify $\dfrac{x^3}{x^5}$.

$$\frac{x^3}{x^5} = \frac{x \cdot x \cdot x}{x \cdot x \cdot x \cdot x \cdot x}$$

$$= \frac{\overset{1}{\cancel{x}} \cdot \overset{1}{\cancel{x}} \cdot \overset{1}{\cancel{x}}}{\underset{1}{\cancel{x}} \cdot \underset{1}{\cancel{x}} \cdot \underset{1}{\cancel{x}} \cdot x \cdot x}$$

$$= \frac{1}{x \cdot x}$$

$$= \frac{1}{x^2}$$

Also, by the quotient rule, $\dfrac{x^3}{x^5} = x^{3-5} = x^{-2}$

Since $\dfrac{x^3}{x^5} = \dfrac{1}{x^2}$ and $\dfrac{x^3}{x^5} = x^{-2}$, thus $x^{-2} = \dfrac{1}{x^2}$.

This can be used to recall the negative exponent rule:

For any nonzero base x, $x^{-n} = \dfrac{1}{x^n}$.

7. Create an example of your own, and use it to show that for any nonzero base x, $x^{-n} = \dfrac{1}{x^n}$.

Scientific Notation

The mass of the Earth is 5,972,000,000,000,000,000,000 metric tons. A really large number like this is hard to work with, so we have a shorthand way of writing really big or really small numbers called scientific notation.

A number in scientific notation is of the form $a \times 10^b$, where $1 \le a < 10$ and b is an integer.

1. Fill in the table with powers of 10.

10^1	10^2	10^3	10^4	10^5	10^6

When 10 has an exponent that is a positive integer, the result is 10 or larger. Positive powers of 10 are used when rewriting a number that is 10 or larger in scientific notation. To convert a number that is 10 or larger to scientific notation, write the decimal point after the first nonzero digit. The exponent of 10 is equal to the number of places the decimal point was moved.

2. Convert to scientific notation.

Standard notation	215,000	2,000,000	1,230,000,000	30,000,000
Scientific notation				

To convert a number from scientific notation to standard notation, multiply the decimal number by the power of 10.

3. Convert each of the following to standard notation.

Scientific notation	2.13×10^4	3.14×10^7	6.7×10^8
Standard notation			

We can also use scientific notation as a shorthand way of writing really small numbers. For example, the diameter of the flu virus is about 0.00000011 meters.

4. Fill in the table with powers of 10.

10^{-1}	10^{-2}	10^{-3}	10^{-4}	10^{-5}	10^{-6}

When 10 has an exponent that is a negative integer, the result is a number between 0 and 1. Negative powers of 10 are used when rewriting a number that is between 0 and 1 in scientific notation. To convert a number that is between 0 and 1 to scientific notation, write the decimal point after the first nonzero digit. The exponent of 10 is negative, and can be determined by the number of places the decimal point was moved.

5. Complete the table.

Standard notation	0.00025		0.0025		0.00000197
Scientific notation		2.97×10^{-4}		9.79×10^{-5}	

We will often multiply or divide numbers in scientific notation. Since multiplication of numbers is commutative, we can multiply in any order.

$$3.62 \times 10^4 \cdot 8.78 \times 10^5 = 3.62 \cdot 8.78 \times 10^4 \cdot 10^5$$
$$= 31.7836 \times 10^9$$
$$= 3.17836 \times 10^{10}$$

To multiply two numbers in scientific notation, first multiply the decimal numbers. Multiply the powers of 10 by using the product rule for exponents: $x^m \cdot x^n$. If the decimal part of the product is 10 or greater you must rewrite the number in scientific notation.

6. Multiply. Express your answer in scientific notation. Round the decimal number to the nearest thousandth.

a. $2.3 \times 10^2 \cdot 6.45 \times 10^3$ 　　　b. $9.7 \times 10^4 \cdot 2.211 \times 10^{-2}$ 　　　c. $7.6 \times 10^5 \cdot 2.6 \times 10^3$

To divide numbers in scientific notation, first divide the decimal numbers. Then divide the powers of 10 using the quotient rule for exponents: $\dfrac{x^m}{x^n} = x^{m-n} \quad (x \neq 0)$.

7. Divide. Express your answer in scientific notation. Round the decimal number to the nearest thousandth.

a. $\left(2.3 \times 10^2\right) \div \left(6.45 \times 10^3\right)$ 　　　b. $\dfrac{9.7 \times 10^4}{2.211 \times 10^{-2}}$ 　　　c. $\left(7.6 \times 10^5\right) \div \left(2.6 \times 10^3\right)$

8. As of December 31, 2018, the total national debt of the U.S. was roughly 21.97 trillion. If the population of the U.S. was 327.1 million at that time, how much is the average government debt per person? (Round to the nearest dollar.)

9. In a state with a population of 7,000,000 people, the average citizen spends $6,000 on housing each year. What is the total spent on housing for the state? (Express your answer in scientific notation.)

Polynomial Expressions

To multiply two polynomials, multiply each term in the first polynomial by each term in the second polynomial and combine any like terms.

$$(x+6)(x+2) = x \cdot x + x \cdot 2 + 6 \cdot x + 6 \cdot 2$$
$$= x^2 + 2x + 6x + 12$$
$$= x^2 + 8x + 12$$

Multiply.

1. $(4x+3)(2x+9)$

2. $(x+6)(3x-8)$

3. $(x-9)(x-13)$

4. $(4x^2+10)(3x^2-16)$

5. $(2x+1)(x^2-7x+10)$

6. $(3x+5)(2x^2-3x-9)$

This approach can also be used to multiply numbers with two (or more) digits.

$$47 \cdot 32 = (40+7)(30+2)$$
$$= 40 \cdot 30 + 40 \cdot 2 + 7 \cdot 30 + 7 \cdot 2$$
$$= 1200 + 80 + 210 + 14$$
$$= 1504$$

Multiply using the same strategy as the strategy for multiplying polynomials.

7. $16 \cdot 75$

8. $84 \cdot 96$

Multiply.

9. $(4x+3)(4x-3)$

10. $(5x-8)(5x+8)$

Notice that the product of two binomials of the form $a+b$ and $a-b$ is equal to a^2-b^2.

Multiply using the fact that $(a+b)(a-b)=a^2-b^2$.

11. $(7x+15)(7x-15)$

12. $(x^3+4y^2)(x^3-4y^2)$

13. Find two binomials whose product is $4x^2-81$.

To square a polynomial, multiply it by itself.
Square the given polynomial.

14. $6x+5$

15. $9x-2$

16. x^2+5x+6

Factoring Trinomials

A trinomial of the form $x^2 + bx + c$ can be factored as the product of two binomials $(x+m)(x+n)$ if $m \cdot n = c$ and $m + n = b$. For certain combinations of b and c, there may be two pairs of m and n that come to mind. Be sure to check that $m \cdot n = c$ and $m + n = b$. This can also be done my multiplying the two factors.

For example, suppose you were trying to factor $x^2 + 5x - 6$, and 2 and 3 immediately pop into your head for m and n. While 2 and 3 do have a sum of 5, their product is 6, not -6. Two other values that can be used are 6 and -1. They have a sum of 5 and a product of -6.

$$x^2 + 5x - 6 = (x+6)(x-1)$$

You can check this by multiplying $(x+6)(x-1)$.

Factor the trinomial.

1. $x^2 - 13x - 30$

2. $x^2 - 13x + 30$

3. $x^2 + 10x - 24$

4. $x^2 + 10x + 24$

5. $x^2 - 26x + 120$

6. $x^2 - 26x - 120$

Factor the trinomial.

7. $x^2 + 0x - 25$

8. $x^2 + 0x - 49$

The two trinomials in exercises 7 and 8 could be rewritten as binomials by omitting the term $0x$. At that point, the binomial is a difference of squares. A **difference of squares** is a binomial that consists of the difference of two perfect squares, $a^2 - b^2$. A difference of squares can be factored using the formula

$$a^2 - b^2 = (a+b)(a-b)$$

Factor the difference of squares. Compare your answer to the factored forms in exercises 7 and 8.

9. $x^2 - 25$

10. $x^2 - 49$

If you forget the formula for factoring a difference of squares, keep in mind that you can always convert the binomial to a trinomial by inserting a term of $+0x$.

If you cannot factor out a common factor from a trinomial whose leading coefficient is not 1, $ax^2 + bx + c$, one strategy is to factor by trial and error. List all the factor pairs of ax^2 and c, and mix and match these pairs in binomials until you find the factored form that would produce bx.

Fill in the missing terms in the factored form of the trinomial.

11. $8x^2 - 46x + 63 = (4x - 9)(\quad - \quad)$

12. $6x^2 + 29x - 120 = (2x + 15)(\quad - \quad)$

13. $6x^2 - 41x - 56 = (6x + \quad)(\quad - 8)$

14. $12x^2 - 25x - 50 = (3x - \quad)(\quad + 5)$

15. $6x^2 + 23x + 20 = (2x + \quad)(3x + \quad)$

12. $5x^2 + 6x - 27 = (5x - \quad)(x + \quad)$

Factor the trinomial. If the three terms share a common factor, factor out the GCF before proceeding.

17. $6x^2 - 18x - 168$

18. $6x^2 + 13x - 5$

19. $7x^2 - 30x + 8$

20. $4x^2 + 88x + 480$

Solving Quadratic Equations

There are four strategies that can be used to solve a quadratic equation: solving by factoring, using the square root property, completing the square, and using the quadratic formula. Certain equations are solved more efficiently with one technique, while other techniques are more efficient for other equations. Being able to select the most efficient technique will save you time and prevent errors.

Solving by Factoring

1. Write the equation in standard form, with all terms collected on one side of the equation set equal to 0 on the other side of the equation.
2. Factor the polynomial.
3. Set each factor equal to 0 and solve.

$$x^2 - 7x + 10 = 0$$
$$(x-2)(x-5) = 0$$

$$x - 2 = 0 \quad \text{or} \quad x - 5 = 0$$
$$x = 2 \quad \text{or} \quad x = 5$$

Whenever possible, factoring is an efficient strategy. Once you have factored the polynomial, finding the solutions is very direct.

If you cannot factor a given polynomial quickly, say in 15 seconds or less, switch to another strategy.

Solving by Using the Square Root Property

This strategy is a wise choice when an equation contains one squared expression and constant terms, but no linear terms.

1. Isolate the squared term on one side of the equation, by collecting all constant terms on the other side of the equation.
2. Take the square root of both sides of the equation. Write $\pm$ in front of the square root of the constant term.
3. Solve the resulting equation.

$$(x-6)^2 - 4 = 21$$
$$(x-6)^2 = 25$$
$$\sqrt{(x-6)^2} = \pm\sqrt{25}$$
$$x - 6 = \pm 5$$
$$x = 6 \pm 5$$
$$x = 11 \quad \text{or} \quad x = 1$$

Although you could have started by rewriting $(x-6)^2$ as $(x-6)(x-6)$, and then worked to get the equation in standard form, using the square root property is far more efficient.

Solving by Completing the Square

1. Isolate the variable terms on one side of the equation, by collecting all constant terms on the other side of the equation.
2. Take half of the coefficient of the linear term, square it, and add it to both sides of the equation.
3. Factor the trinomial as a perfect square.
4. Solve by using the square root property.

$$x^2 - 8x - 10 = 23$$
$$x^2 - 8x = 33$$
$$x^2 - 8x + 16 = 33 + 16$$
$$(x-4)^2 = 49$$
$$\sqrt{(x-4)^2} = \pm\sqrt{49}$$
$$x - 4 = \pm 7$$
$$x = 4 \pm 7$$
$$x = 11 \quad \text{or} \quad x = -3$$

Under the right conditions, this technique can be more efficient than solving by using the quadratic formula. If the coefficient of the squared term is 1, and the coefficient of the linear term is even, then completing the square can be a great choice. If the leading coefficient is not equal to 1 or if the linear term has an odd coefficient, it will be more efficient to try another method like using the quadratic formula.

Solving by Using the Quadratic Formula

1. Write the equation in standard form, $ax^2 + bx + c = 0$, with all terms collected on one side of the equation set equal to 0 on the other side of the equation.
2. Identify a, b, and c.
3. Substitute into the quadratic formula and simplify.

$$x^2 - 8x - 10 = 23$$
$$x^2 - 8x - 33 = 0$$
$$x = \frac{-(-8) \pm \sqrt{(-8)^2 - 4(1)(-33)}}{2(1)}$$
$$x = \frac{8 \pm \sqrt{196}}{2}$$
$$x = \frac{8 \pm 14}{2}$$
$$x = \frac{8 + 14}{2} = 11 \quad \text{or} \quad x = \frac{8 - 14}{2} = -3$$

Although the quadratic formula can be used to solve any quadratic equation, it is not always the most efficient way to go. You must be careful when performing the arithmetic. Notice that this equation could have been solved by factoring, and it would likely take far less time to solve it that way.

Solve by the most efficient technique.

1. $x^2 - 3x - 108 = 0$

2. $2x^2 - 9x + 10 = 0$

3. $(x+2)^2 + 10 = -22$

4. $x^2 - 10x = 23$

5. $x^2 - 7x + 2 = 0$

6. $x^2 + 44 = -20$

7. $3x^2 - 10x + 7 = 0$

8. $x^2 + 12x + 36 = 0$

9. $x^2 + x - 210 = 0$

10. $10x^2 + 9x = 7$

Percent Applications

Many consumer applications involve percentages. Some, such as sales tax, markup, commission, and tips, are added onto an amount. Others, such as discount, are subtracted from an amount. Occasionally, two (or more) quantities are added to or subtracted from an amount.

Sales tax is a payment made to the government on most things that are bought or sold. Sales tax is usually a percentage of the total value of the items. For example, if we buy $60 of ice cream for a party, and the tax rate is 8%, then we would pay a total of $0.08(\$60)$ or $4.80 in sales tax. The total price would be $\$60 + \$4.80$, or $64.80.

1: Calculate the sales tax and the total price. (Round to the nearest cent.)

Pre-tax price and tax rate	$45.00 , 6%	$45.00, 8%	$50.00, 5.5%	$62, 8.25%
Sales tax				
Total price				

A tip is left for a server in a restaurant for providing good service. The tip is usually computed as a percentage (usually 15% - 20%) of the total. If the bill was $57.52, a 15% tip would be $0.15(\$57.52)$ or $8.63, rounded to the nearest cent. Adding the tip to the bill produces a total bill of $\$57.52 + \$8.63$, or $66.15.

2. Calculate the tip and the total bill. (Round to the nearest cent.)

Sub total	$41.99	$38.27	$8.99
Tip percentage	15%	20%	20%
Tip			
Total bill			

Adding sales tax to a purchase or a tip to a bill can be thought of as a percent increase. Another application of percent increase involves a store marking up their prices. Stores typically markup their prices by a certain percentage over what the item cost them to acquire or make.

Percent decrease refers to a quantity that is lowered by a certain percentage of the original quantity. One application involving percent decrease involves sales discounts. If a $60 item is on sale at a 25% discount, the amount of the discount is $0.25($60)$ or $15. The discount is then subtracted from the original price, so the sale price of this item is $60 - 15, or $45.

3. Calculate the percent increase or decrease and the resulting price. (Round to the nearest cent.)

Initial price	$32	$45.00	$45.00	$50.00
Percent change	23% increase	12% increase	12% decrease	13% increase
Final price				

A commission is a bonus paid to someone for being successful. Usually, a commission is a percentage of your total sales. For example, a real estate agent typically gets a 3% commission based upon the selling price of the house. If a house sold for $175,000, the typical realtor commission would be $0.03($175,000)$ or $5250. The seller of the house pays the commission out of the sales price. In this example, there would be $175,000 - 5250 or $169,750 remaining for the seller.

4. Calculate the commission, and determine how much of the sales total remains for the seller. (Round to the nearest cent.)

Sales total	$155,000	$155	$173,000
Commission rate	4.5%	4.5%	1.2%
Commission			
Amount Remaining for the Seller			

Often, two or more of these increases or decreases are applied in the same problem. The increases must be applied successively, one after the other, and not combined into one percent. For example, if a store marks its prices up by 50% in a state that charges 9% sales tax, you must first increase the price by 50% and then the new price by 9%. You cannot simply increase the price by 59%.

5. Rob went to buy a new phone. The original cost of $825 had been marked up by 20%, and Rob had a coupon for 30% off. The sales tax rate is 8.25%.
a. What is the price of the phone after the 20% markup is applied?

b. What is the price of the phone after Rob applies his 30% off coupon?

c. What is the price of the phone after the 8.25% sales tax is applied? (Round to the nearest cent.)

6. Alexis is shopping for clothes. She picks some items totaling $120. She has two coupons, a $10 off and a 15% off. The store says she must first use the $10 off coupon. If the tax rate is 7%, what is her total cost?

7. Paige is selling two different houses. One is priced $150,000 and the other is priced $110,000. If the price of the more expensive house drops 8% while the price of the other house drops 5% before they sell, what is her commission if she makes a 4% commission on the sale?

8. Jacob is selling new cars. He is trying to sell three more cars before the end of the month. Each car has a sale price of $50,000. He has two options to help make the sales: he can either lower the price by 30% and keep his commission at 3%, or he can lower the lower the price by 10% and also lower his commission to 2% If he sells all three cars, how much total commission would he make in each case?

Simple Interest

The formula for computing simple interest is $I = Prt$ where I is the interest, P is the principal, r is the interest rate, and t is the time in years.

1. Compute how much interest each account would earn in one year for the given principal and interest rate.

Principal (P)	a. $100	b. $500	c. $200	d. $100	e. $200
Interest rate (r)	3%	3%	3%	6%	6%
Interest (I)					

2 Find the interest earned in one year for the given principal and interest rate, and find the balance in the account after one year.

Principal (P)	a. $400	b. $500	c. $300	d. $850	e. $225
Interest rate (r)	3%	7%	2%	6%	4%
Interest (I)					
Balance after one year					

3 Find the interest earned for the given principal, interest rate, and time period, and find the balance in the account after the given time period.

Principal (P)	a. $400	b. $500	c. $2300	d. $400	e. $400
Interest rate (r)	3%	7%	3%	3%	3%
Time (t)	5 years	5 years	5 years	9 years	18 years
Interest (I)					
Balance					

In exercise 3 we added the interest earned to the principal to find the total amount of money in the account.

$$P + I = P + Prt$$
$$= P(1 + rt)$$

This is why the formula is usually given as $A = P(1 + rt)$ where A is the balance after time t, P is the principal, r is the interest rate, and t is the number of periods, or time.

4. Compute the balance after the given time period.

Principal (*P*)	a. $1000	b. $2000	c. $1000	d. $2000	e. $4000
Interest rate (*r*)	4%	4%	5%	5%	10%
Time (*t*)	25 years	25 years	20 years	20 years	10 years
Interest (*I*)					
Balance					

5. What do you notice about each of the total amounts in exercise 4?

6. For simple interest, does there appear to be a rule for determining the doubling time? (i.e. the time it takes an initial investment to double)

Compound Interest

If interest is compounded n times per year, the balance A after t years can be found from the formula

$A = P\left(1 + \dfrac{r}{n}\right)^{nt}$, where P is the principal and r is the annual interest rate. To determine the amount of interest earned, subtract the principal from the balance after t years: $A - P$.

1. Find the total amount of interest earned by a deposit of $1000 under the given conditions, if interest is compounded annually $(n = 1)$.

a. 3% annual interest, 5 years   b. 6% annual interest, 5 years   c. 6% annual interest, 10 years

2. Compare your answers from parts a and b in exercise 1. Does doubling the interest rate double the interest earned?

3. Compare your answers from parts b and c in exercise 1. Does doubling the time period double the interest earned?

4. Find the total amount of interest earned by a deposit of $1000 in 5 years under the given conditions.

a. 9% annual interest compounded quarterly $(n = 4)$

b. 9% annual interest compounded monthly $(n = 12)$

c. 9% annual interest compounded semi-annually $(n = 2)$

5. Look at your results from exercise 4. Would you expect the interest earned to be less than or greater than those amounts if $1000 was invested at an annual interest rate of 9% compounded daily $(n = 365)$ for 5 years? Explain your answer.

The doubling time (the amount of time for the balance to double) for simple interest can be found using the formula $t = \dfrac{1}{r}$. Now let's try to find a similar relationship for compound interest.

6. Determine the balance after time t under the given conditions.
a. $400 invested at 4% annual interest, compounded annually, for 18 years.

b. $3000 invested at 6% annual interest, compounded semi-annually, for 12 years.

c. $500 invested at 8% annual interest, compounded quarterly, for 9 years.

d. $2400 invested at 9% annual interest, compounded monthly, for 8 years.

7. Notice that in parts a-d of exercise 6, the principal roughly doubled during the term of the investment. For parts a-d of exercise 6, what is the product of the interest rate and time?

Many financial advisors use the "rule of 72" to determine how long it will take for an investment to double. To apply the rule of 72, divide 72 by the interest rate to estimate the doubling time in years.

Order of Operations and Function Notation: $f(x + h)$ versus $f(x) + h$

1. For $f(x) = x^2$, evaluate each of the following expressions in order, a through f.

a. $f(2 + 3)$	b. $f(2) + 3$
c. $f(4 \times 3)$	d. $f(4) \times 3$
e. $f(3 \times 2 + 1)$	f. $3f(2) + 1$

2. In evaluating each of the expressions above, several operations were performed. List the operations, add, multiply and/or square, in the order in which they were performed.

a.	b.
c.	d.
e.	f.

3. Match each variable expression in the table on the left with the equivalent expression in function notation, where $f(x) = x^2$.

Variable Expressions	
A.	$3x^2$
B.	$(3x)^2$
C.	$3(x + 3)^2$
D.	$3x^2 + 3$
E.	$3(x + 3)^2 + 3$
F.	$(3x + 3)^2 + 3$
G.	$(3x + 3)^2$

Function Notation	Letter of Equivalent Variable Expression
$f(3x + 3)$	
$f(3x + 3) + 3$	
$3f(x + 3) + 3$	
$3f(x)$	
$3f(x + 3)$	
$f(3x)$	
$3f(x) + 3$	

4. Let $f(x) = x^2$. Rewrite each expression using function notation. Informally, instead of writing, "$(expression)^2$," write, "$f(expression)$."

   Example: The expression $5(x + 1)^2 - 3$ can be written in function notation as $5f(x + 1) - 3$.

a. $4x^2$	b. $x^2 + 5$
c. $-2x^2 + 4$	d. $(x - 2)^2$
e. $7(x + 9)^2$	f. $(x - 3)^2 + 1$
g. $2(x - 5)^2 - 6$	h. $-(2x)^2 + 3$

5. Let $f(x) = x^2$. Evaluate the expressions across each row at the given values of a and h.

	$f(a + h)$	$f(a) + h$	$f(a) + f(h)$
$a = 1$ $h = 2$			
$a = 3$ $h = 7$			
$a = -1$ $h = 7$			
$a = 5$ $h = -2$			

The examples in the table above support the fact that, for a general function f and for real numbers a and h,

$$f(a + h) \neq f(a) + h \neq f(a) + f(h)$$

$f(a + h) \neq f(a) + h$ is a symbolic way of saying that order matters. In the expression $f(a + h)$, the addition of h occurs BEFORE the function f is applied; whereas, in the expression $f(a) + h$, the addition of h occurs AFTER the function f is applied.

Warning: If read aloud, the expressions, $f(a + h)$ and $f(a) + h$ sound identical: "f-of-a-plus-h;" however, they mean different things! For clarity, insert a pause for emphasis:

$f(a + h)$ reads, "f-of-PAUSE-a-plus-h, whereas $f(a) + h$ reads, "f-of-a-PAUSE-plus h."

$f(a + h) \neq f(a) + f(h)$ indicates that the function f does not "distribute." Applying f to the sum of a and h is different than applying f to both a and h then adding the results.

6. Let $f(x) = x^2$. Evaluate and simplify the expressions, working across each row.

	$f(ax)$	$af(x)$
$a = 5$		
$a = -2$		
$a = \frac{1}{2}$		
$a = -\frac{3}{4}$		

The examples in the table above support the fact that, for a general function f and for real number a,

$$f(ax) \neq af(x)$$

In other words, you cannot factor out constants from the parentheses in function notation. Multiplying then applying a function is not the same as applying a function then multiplying.

7. Read the following three responses to the instructions, "Evaluate and simplify the expression $f(x + h)$ for the function $f(x) = x^2 - x$."

    Response 1:
    (1) Start with the expression for $f(x)$: $x^2 - x$
    (2) Place parentheses around each x in the expression: $(x)^2 - (x)$
    (3) Replace x with $x + h$ throughout the expression: $(x + h)^2 - (x + h)$
    (4) Simplify: $x^2 + 2xh + h^2 - x - h$

    Response 2:
    (1) Start with the expression for $f(x)$: $x^2 - x$
    (2) Replace x with $x + h$: $(x + h)^2 - x + h$
    (3) Simplify: $x^2 + 2xh + h^2 - x$

    Response 3:
    (1) Start with the expression for $f(x)$: $x^2 - x$
    (2) Instead of $f(x)$, we want $f(x + h)$, so add h: $x^2 - x + h$

    For each of Responses 1, 2 and 3, state that it is correct, or indicate the line number where an error was made, explain what the error was and fix it.

    Response 1:

    Response 2:

    Response 3:

8. Simplify the expression $f(x + h)$ for each function f.

    a. $f(x) = x^2 + 2x$

    b. $f(x) = 3x^2 - x$

    c. $f(x) = -x^2 + 4x - 2$

9. For each function, simplify the expression $f(x + h)$ then find the difference between $f(x + h)$ and $f(x)$.

    a. $f(x) = x^2 - 3$

    b. $f(x) = x^2 - 2x$

Linear Functions

A linear function is a function that has a constant rate of change. For example, if you work a few hours a week and are paid an hourly wage, that is an example of a linear function because for each additional hour you work you gain a set amount of money. To determine if a set of ordered pairs (x, y) represents a function, determine if the rate of change (or slope) remains constant for each pair of ordered pairs.

Consider the table of ordered pairs.

x	1	3	9
y	2	4	10

This represents a linear function.

The rate of change for the ordered pairs $(1,2)$ and $(3,4)$ is computed $m = \dfrac{4-2}{3-1} = \dfrac{2}{2} = 1$.

The rate of change for the ordered pairs $(1,2)$ and $(9,10)$ is computed $m = \dfrac{10-2}{9-1} = \dfrac{8}{8} = 1$.

The rate of change is also the same for the ordered pairs $(3,4)$ and $(9,10)$.

Now consider the following table of ordered pairs.

x	1	4	8
y	2	6	10

This does not represent a linear function because the rate of change is not constant.

The rate of change for the first two ordered pairs is $m = \dfrac{6-2}{4-1} = \dfrac{4}{3}$.

The rate of change for the last two ordered pairs is $m = \dfrac{10-6}{8-4} = \dfrac{4}{4} = 1$.

1. Determine whether the function is linear.

a.

x	y
2	5
4	7
5	8

b.

x	y
2	4
4	8
5	11

c.

x	y
1	5
2	7
3	9

d.

x	y
2	3
5	5
6	7

A linear function is usually expressed in function notation. A linear function is of the form $f(x) = mx + b$, and its graph is a line.

- The y-intercept of the line is the point $(0,b)$ because substituting 0 for x produces an output of b:
$$f(0) = m \cdot 0 + b = b.$$
- The slope of the line is m because each increase in x by 1 produces an increase in the function by m.

2. Find the slope and y-intercept.

a. $f(x) = \dfrac{4}{3}x - 7$　　　b. $f(x) = 9 - 2x$　　　c. $f(x) = 5x + 2$　　　d. $f(x) = 3x$

3. Consider the four graphs of linear functions, A – D.

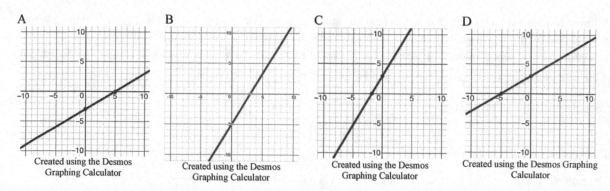

A

Created using the Desmos
Graphing Calculator

B

Created using the Desmos
Graphing Calculator

C

Created using the Desmos
Graphing Calculator

D

Created using the Desmos Graphing
Calculator

a. Which graph has a slope of $\dfrac{5}{3}$ and a y-intercept of $(0,3)$?

b. Which graph has a slope of $\dfrac{5}{3}$ and an x-intercept of $(3,0)$?

c. Which graph has a slope of $\dfrac{3}{5}$ and a y-intercept of $(0,3)$?

d. Which graph has a slope of $\dfrac{3}{5}$ and an x-intercept of $(5,0)$?

4. Lydia is starting a new job as a nurse. She will get a $5000 signing bonus, and then will be paid $35 per hour. Write a linear function $f(x) = mx + b$ for the amount of her first paycheck if x is the number of hours she works. Use the model to determine how large her paycheck will be if she works 34 hours the first week. (Assume the signing bonus is included in her paycheck.) Use an appropriate scale and graph the function.

5. In an 1897 copy of the *American Naturalist*, Dr. Amos Dolbear published an article where he mentioned his observations that at 60 degrees Fahrenheit, the crickets chirped at a rate of 80 per minute, and at 70 degrees, they chirped at a rate 120 per minute.

a. Use this information to find a linear function $f(x) = mx + b$ for the chirps/minute of a cricket as a function of temperature x (in degrees Fahrenheit).

b. What is the x-intercept for this function, and what does it mean? What does this suggest about the domain of this function?

Comparing and Contrasting Data Displays

Consider the following data displays when answering all subsequent questions.

line graph   bar graph   frequency distribution
relative frequency distribution   histogram   circle graph
scatterplot

1. From which data displays is it possible to determine the number of data values in the set? For each of those, describe how that number can be determined.

2. Which data displays can be used for data that is categorical?

3. Which data displays have an axis that measures frequency?

4. Which data displays can be used for a data set if some values may fall into more than one category?

5. Which data displays have at least one axis that gives a numerical measure of a quantity other than a frequency?

6. Which data displays consist of bars that may be arranged in any order; in particular, from tallest to shortest?

7. Which data displays can be used when only one variable is measured for each member or object involved?

8. Which data displays can be used to show how many data values fall into specific (numeric) intervals?

State the best data display for each purpose.

9. Show how the sizes of each category compare to the size of the whole collection of categories.

10. Show how one quantity changes over time.

11. Show how the sizes of the categories compare to each other.

12. Show the shape of the data.

13. Explore a relationship between two variables for each member or object involved.

14. Use percents instead of counts for large data sets of numerical measures.

For each situation and data display, state what each component of the display represents. Give all relevant information, such as numerical value(s) and units, whenever applicable.

15. A circle graph is used to represent a set of data collected from 100 online purchasers about which online retailer they prefer.

    The circle:

    The sectors:

16. A histogram is used to display the total cost for tuition and books for 1200 college students this semester.

    The horizontal axis:

    The vertical axis:

    The height of each bar:

17. A scatter plot is used to display the number of milligrams of artificial sweetener 36 individuals consumed in one day and their calorie intake that day.

    The horizontal axis:

    The vertical axis:

    A point:

Comparing Measures of Center for Distributions of Different Shapes

The students of a statistics class collected data about themselves.

1. Each student reported his/her age. The histogram illustrates the results.

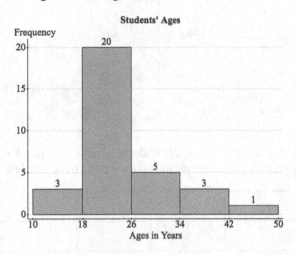

a. How many students are in this class? Show a calculation to support your answer.

b. Can the following list of ages, in years, be the data that was collected by these students? If not, edit the list so that it can be.

| 16 | 17 | 17 | 18 | 18 | 18 | 19 | 19 | 19 | 19 | 19 | 20 | 20 | 20 | 21 | 21 |
| 21 | 22 | 23 | 23 | 24 | 24 | 25 | 27 | 30 | 31 | 31 | 33 | 35 | 38 | 40 | 46 |

c. If the histogram illustrates the set of ages listed above, do the numbers labeled on the horizontal axis of the histogram fall into the bin on their left or on their right? How can you tell?

d. Using the set of ages listed above, compute each of the following measures of center: mean, median, mode, midrange.

e. Find the approximate locations of the mean and median on the horizontal axis of the histogram. Label them as "M" and "med," respectively.

2. Each student reported the number of units she/he is enrolled in, as illustrated in this histogram.

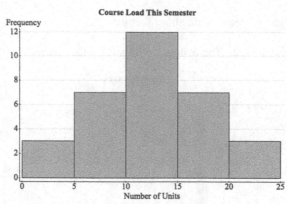

a. Create a list of possible numbers of units the students in this class are taking this semester. Assume that each number labeled on the horizontal axis falls into the bin on its left.

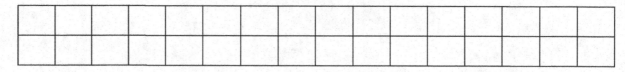

b. Compute the mean and median of your data set. Find the approximate locations of the mean and median on the horizontal axis of the histogram. Label them as "M" and "med," respectively.

3. The histograms below illustrate two other data sets collected by and about these students. For each data set, compute the mean and median, and label those values on the horizontal axis of the histogram for that set.

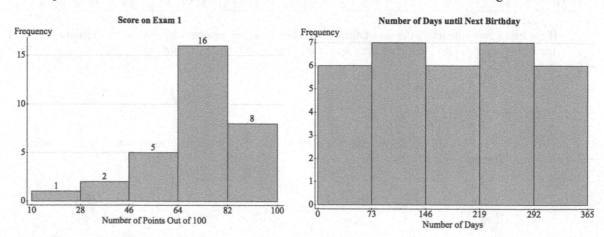

Data Set for Score on Exam 1

11	30	42	48	52	57	60	63	65	65	67	67	69	70	70	70
76	76	77	77	79	80	80	80	83	83	85	87	89	90	94	96

Data Set for Number of Days Until Next Birthday

10	29	43	51	62	69	78	90	99	108	123	142	143	154	162	177
193	202	213	229	240	248	256	273	280	290	299	303	341	350	352	360

4. For each of the four sets of data, compute the difference between the mean and the median. Then, find the relative difference, rounded to the nearest tenth of a percent. (Hint: Find the ratio of the difference between the mean and median to the median.)

Data Set	Difference between mean and median	Relative difference between mean and median
Age		
Number of Units		
Score on Exam 1		
Days Until Birthday		

a. Give the title(s) of the data set(s) for which the mean is greater than the median by more than 3%.

b. Give the title(s) of the data set(s) above for which the mean is less than the median by more than 3%.

c. Give the title(s) of the data set(s) above for which the mean and median are approximately equal (the mean is within 3% of the median).

d. The names for the shapes of the four histograms, in order, are: skewed right, normal, skewed left and uniform. Fill in the blank of the following generalizations with one of the words, "mean," or "median," or one of the shapes named above.

If the shape of the histogram for a data set is skewed left, the _____ is greater than the

_____.

If the shape of the histogram for a data set is skewed right, the _____ is greater than the

_____.

If the shape of the histogram is _____ or _____, the mean and median are

(approximately) equal.

Optional Class Activity

Poll the students in your class to determine the number of units each is enrolled in this semester. By show of hands, count the number of students in your class who fall into each class to complete the frequency distribution below.

5.	7. Draw the histogram for the frequency distribution.

5.

Number of Units	Frequency
1 − 5	
6 − 10	
11 − 15	
16 − 20	
21 − 25	

6. Predict: Based only on the frequencies in the frequency distribution, what shape do you think the histogram for this data set will be?

8. Predict: Based only on the shape of the histogram, which is larger, the mean or the median?

9. Poll the class again; however, this time, record the number of units each student is taking this semester.

10. Compute the mean and the median. Which is larger? By what percent is the mean different than the median?

11. Repeat this activity by polling the class to determine the number of days until each student's next birthday. For the frequency distribution, use the same classes as the histogram from Question 3.

Counting With and Without Replacement

1. Suppose that five-letter "words" are to be constructed using only the letters, A, B, C, D and E.

    a. **If repetitions are not allowed**, find the number of five-letter words that can be constructed by completing the table and applying the Multiplication Principle.

Position	1st	2nd	3rd	4th	5th
Number of Choices					

    b. **If repetitions are allowed**, find the number of five-letter words that can be constructed by completing the table and applying the Multiplication Principle.

Position	1st	2nd	3rd	4th	5th
Number of Choices					

2. The number of ways to order five items selected from a set of five distinct items depends on whether or not repetitions are allowed. Fill in each blank to complete an expression for the number of arrangements in each case.

    a. Without repetitions: $\boxed{}!$

    b. With repetitions: $\boxed{}^{\boxed{}}$

3. A major scale in music consists of 8 notes. Give an expression for the number of musical phrases that can made from a major scale in each case and calculate that value.

    a. the number of 8-note phrases if a note may be used more than once.

    b. the number of 8-note phrases if each note may be only once.

    c. the number of 6-note phrases if a note may be used more than once. (The musical phrase corresponding to the words "Happy Birthday to You" is such a phrase.)

4. A coin is to be flipped 20 times. Find an expression, and calculate its value, for the number of 20-letter sequences, selecting from the letters H (for heads) and T (for tails), that are possible.

Factorials are used to determine the number of possible arrangements for a number of distinct items. For example, if there are seven people, in how many different ways can they sit in seven chairs in a row? For the first seat there are seven people who could sit there, but then for the next seat there are only six choices (one person is already seated). Continuing in the same fashion, the next seat only has five choices, and so forth. We then multiply the number of options for each seat. The total number of arrangements is $7 \cdot 6 \cdot 5 \cdot 4 \cdot 3 \cdot 2 \cdot 1$ or $7!$, which is equal to 5040.

What if we had seven people, but only three seats? This problem can be solved with multiplication. We have seven choices for the first seat, six choices for the second seat, and five choices for the third seat. The number of possible arrangements is $7 \cdot 6 \cdot 5$ or 210.

5. For the given number of people and seats, find the number of different ways to seat people.

People	8	6	3	6	4	10	40
Seats	8	6	3	4	2	7	2
Number of ways							

How many ways are there to rearrange the letters in ISOGRAM? The letters in isogram are all distinct, so this is similar to the seating example above – we are trying to decide how many letters could be first, how many could be second, and so on. Seven letters, seven places: $7! = 5040$

6. How many ways are there to rearrange the letters in each of the following words?

Word	campground	countryside	count
Number of rearrangements			

How many ways are there to rearrange the letters in HELLO? This is different, because of the repeated L's. Does switching the L's give a different rearrangement? In mathematics, we say no, as you cannot tell the difference. Hence, we usually ask "How many distinguishable rearrangements of the letters in HELLO are there?" There are two L's, and $2! = 2$ ways to arrange the two L's, so we can divide out by the ways that will be counted the same.

$$\frac{5!}{2!} = \frac{5 \cdot 4 \cdot 3 \cdot 2 \cdot 1}{2 \cdot 1} = 5 \cdot 4 \cdot 3 = 60$$

What about BASEBALL? There are two B's, two A's, and two L's.

$$\frac{8!}{2!2!2!} = \frac{5040}{2 \cdot 2 \cdot 2} = 630$$

7. How many distinguishable rearrangements of the following words are there?

Word	apples	queue	mathematics
Number of rearrangements			

Percentiles and Probabilities from Pictures

1. The heights, in inches, of 200 American males, ages 20 and older, were measured. The frequency table and histogram summarize and illustrate the results.

Height (inches)	Frequency
55 − 59	5
60 − 64	27
65 − 69	68
70 − 74	68
75 − 79	27
80 − 84	5

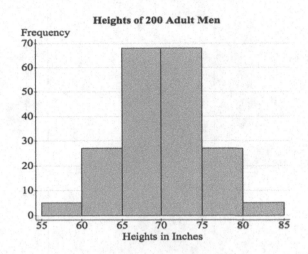

Heights of 200 Adult Men

a. What percent of these men's heights are in the middle two classes?

b. What percent of these men's heights are in the middle four classes?

c. Find the cut-off height for the shortest 2.5% of these men. In other words, if one these 200 men is in the shortest 2.5%, then his height must be less than _____ inches.

d. Find the cut-off height for the tallest 16% of these men. In other words, if one these 200 men is in the tallest 16% of them, then his height must be at least _____ inches.

e. How many of the men's heights, x, satisfy the inequality, $60 \leq x < 75$?

f. Using diagonal lines, shade a portion of the histogram to illustrate the number found in part e.

g. If one man is randomly selected from these 200 men, what is the probability that his height, x, satisfies the inequality, $60 \leq x < 75$?

h. Compute the total area of the shaded bars, and the total area of all the bars in the histogram, disregarding the fact that the horizontal axis is measured in inches.

i. Find the ratio of the shaded portion to the entire histogram equal to the probability found in part f?

2. A value, x, is randomly selected from the data set whose histogram is given. Calculate the stated probability. Round to three decimal places. Shade the corresponding portion of the histogram whose ratio to the entire histogram is the same as the probability.

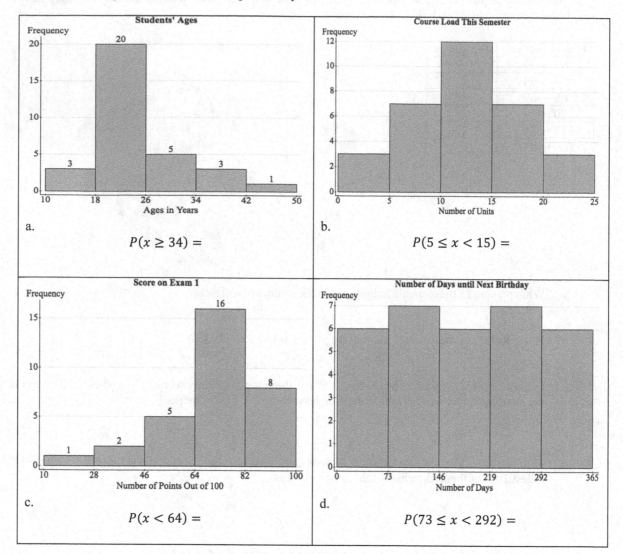

a.
$$P(x \geq 34) =$$

b.
$$P(5 \leq x < 15) =$$

c.
$$P(x < 64) =$$

d.
$$P(73 \leq x < 292) =$$

3. One score on Exam 1 is randomly selected. Use the histogram above to compute $P(x < 82)$ in two different ways.
   a. Use addition.
   b. Use subtraction.

4. The histogram for a data set has first lower class limit of 10 and class with of 5. Label the horizontal axis on its histogram below, accordingly. Is $P(x < 25)$ greater than, less than or equal to 0.5? Explain.

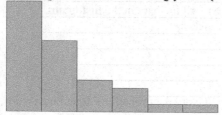

Venn Diagrams

Pictures of sets and set relationships are often helpful in understanding the concepts. We will use Venn diagrams to help us with these concepts. A Venn diagram consists of a rectangle representing the universal set U, and a circle for each set being investigated. When working with two sets, the circles should overlap.

Suppose we are looking at the relationship of two sets A and B. Here is the Venn diagram for examining the relationship between those two sets.

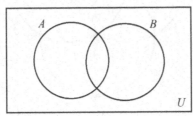

The union of A and B, $A \cup B$, is made up of items that are in A or in B or in both.

The intersection of A and B, $A \cap B$, is made up of items that are in both A and B.

The complement of A, denoted A', is made up of items that are not in A.

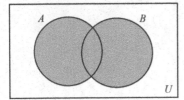
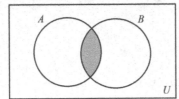
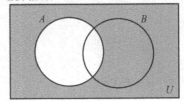

1: For two sets A and B, match the set relationship with the correct Venn diagram.

a. U b. A c. $A \cap B$ d. $A \cup B$ e. B' f. $A' \cap B$

g. $A' \cup B$ h. $A \cup B'$ i. A' j. B k. $(A \cup B)'$ l. $\varnothing$

1.
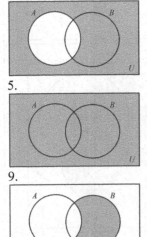

2.

3.

4.

5.

6.
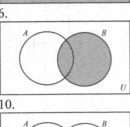

7.
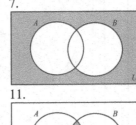

8.

9.
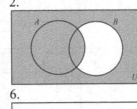

10.

11.
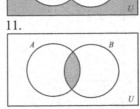

12.
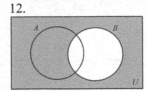

Suppose we are looking at the relationship of three sets *A*, *B*, and *C*. Here is the Venn diagram for examining the relationship between those three sets.

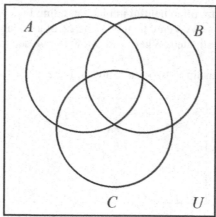

2. For three sets *A*, *B*, and *C*, match the set relationship with the correct Venn diagram.

a. $(A \cup B) \cap C$ b. $(A \cap B) \cup C$ c. $A \cup (B \cap C)$

d. $A \cap (B \cup C)$ e. $A \cup B \cup C$ f. $A \cap B \cap C$

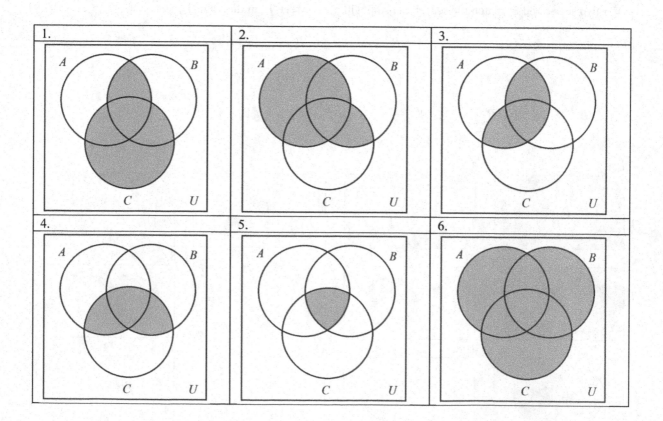

Euler Diagrams and Valid Arguments

An **Euler diagram** is similar to a Venn diagram in that it is used to show the relationship between sets. While a Venn diagram shows all possible relationships between two (or more) sets, an Euler diagram only shows the relationships that actually exist. For example, if *A* and *B* have no elements in common, an Euler diagram shows two circles that do not intersect while a Venn diagram would show that empty intersection.

Here are four Euler diagrams that show how two sets *A* and *B* relate to each other.

All *A* are *B*.   No *A* are *B*.

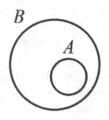

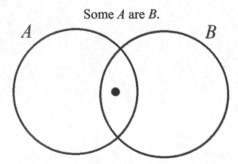

Some *A* are *B*.   Some *A* are not *B*.

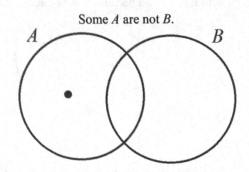

1. Draw an Euler diagram to show the relationship between the sets.

a. *A*: Set of math professors, *B*: Set of professors   b. *A*: Set of birds, *B*: Set of fish

c. *A*: Set of college freshmen,
B: Set of college students

d. *A*: Set of college students,
B: Set of elementary school children,
C: Set of college freshmen

Euler diagrams can be used to help determine if an argument is valid.

Consider this argument: All college students (*A*) take classes (*B*). Rob is a college student. Therefore, Rob takes classes. Start with an Euler diagram that shows that all *A* are *B*. Since Rob is a member of set *A*, he must also be a member of set *B*. The argument is a valid argument.

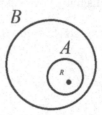

In a deck of playing cards, all diamonds are red cards. A red card is drawn from the deck. Therefore, the card is a diamond. Let *A* be the set of diamonds and *B* be the set of red cards. Start with an Euler diagram that shows that all *A* are *B*. Since it is possible that some members of *B* are not members of *A*, namely cards that are hearts, this argument is not valid because it is not necessarily true.

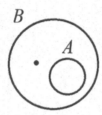

2. For each argument, use an Euler diagram to determine if the argument is valid.

a. All math professors took calculus. Mark took calculus. Therefore, Mark is a math professor.

b. All math professors took calculus. Mark is a math professor. Therefore, Mark took calculus.

c. All poodles are dogs. Spot is a dog. Therefore, Spot is a poodle.

d. All vegetarians like to eat vegetables. Rob likes to eat vegetables. Therefore, Rob is a vegetarian.

e. All Red Sox fans are baseball fans. George is a Red Sox fan. Therefore, George is a baseball fan.

Is That a Big Number?

One place where ratios and rates are often seen is in the reporting of the frequency of various bad events, like deaths from certain illnesses. To help people get a handle on how big a number can be, they are often reported as a number of people per week or some other unit of time.

1. In 2018, according to the national cancer institute of America, almost 70 people per hour died of cancer in the United States. According to the Center for Disease Control (CDC) in Atlanta, GA, every 52 seconds someone in the US died of heart disease in 2018. Which illness had the most total deaths in 2018?

2. In 2018, the estimated population of the US was 310 million people. What percent of the US population died of cancer in 2018? (Round to the nearest tenth of a percent.)

3. Fill in the following chart.

a. Hours in a year	b. Minutes in a year	c. Seconds in a day	d. Seconds in a week	e. Seconds in a year

4. How long in days would a million seconds be? (Round to the nearest tenth of a day.)

5. How long in days and years would a billion seconds be? (Round to the nearest day.)

6. How long in years would a trillion seconds be? (Round to the nearest year.)

7. For the fiscal year 2019 (October 1, 2018 to September 30, 2019) the US federal budget is $4.4 trillion. If politicians are arguing over a program that will cost $12 million, what percent of the federal budget is that? (Round to the nearest ten-thousandth of a percent.)

8. What percent of the federal budget would a program that costs $12 billion be? (Round to the nearest ten-thousandth of a percent.)

9. The volume of 1 million pennies is approximately 20 cubic feet. How many pennies would it take to replace the desk / table you are sitting at? (Include the space under the desk/table..)

10. Roughly how many pennies would it take to fill your current classroom?

11. A million pennies laid out flat on the ground would cover 3921 square feet. Roughly how many pennies would it take to cover your school's campus in a layer of pennies?

Applying Linear Equations, Inequalities and Formulas: Volume Variations

An aluminum can has a diameter and height of approximately 2.6 inches and 4.5 inches, respectively.

1. Compute the volume of this can. Round to the nearest hundredth of a cubic inch.

2. Use the fact that $1 \text{ in}^3 \approx 0.55$ fluid ounces to compute the liquid capacity of this can. Round to the nearest hundredth of a fluid ounce.

This can contains 12 ounces of soda. The total cost to purchase each 12-ounce can of soda includes a deposit of $0.05, which will be refunded when the can is recycled. The total cost of a 6-pack of 12-ounce sodas, including the recycling deposit for each can, is $13.73.

3. Write an equation that gives the total cost of a 6-pack of 12-ounce sodas in terms of x, the cost, in dollars, for one ounce of soda.

4. Solve the equation.

5. A 2 liter bottle contains approximately 68 fluid ounces, but the recycling deposit is 10¢. Find the total cost for a 2 liter bottle of the soda, including the recycling deposit, if the unit price of the soda is the same regardless of how it is packaged.

6. A pallet of 12-packs of 12-ounce cans of soda consists of twenty 12-packs per tier, stacked 10 tiers high. How many cans of soda are on a pallet?

7. How many ounces of soda are on the pallet?

8. Find the total cost of all the soda on the pallet, without recycling fees.

Slope and Staircases

In the United States, most places specify a ratio for how much each step in a set of stairs can go up compared to how wide or deep each step needs to be. A general rule of thumb for homes is two times the step riser height plus the tread width be between 24 and 25 inches, inclusive.

1. Determine which of the following patterns would fit this rule of thumb.

Step height: tread width	a. 7 in: 12 in	b. 8 in: 9 in	c. 7 in: 11 in	d. 6 in: 12 in
Rule of thumb total $2(\text{step height}) + (\text{tread width})$				
Passes rule of thumb?				

2. Compute the slopes of each step pattern from exercise 1.

a. 7 in: 12 in   b. 8 in: 9 in   c. 7 in: 11 in   d. 6 in: 12 in

3. Suppose that the step height and tread width must be integers.
For the given tread width, find the step height that satisfies the rule of thumb that two times the step riser height plus the tread width be between 24 and 25 inches, inclusive.

a. Tread width: 10 in   b. Tread width: 11 in

c. Tread width: 12 in   d. Tread width: 13 in

4. If a town's building code specifies a minimal width of ten inches for the tread width and a minimal height of four inches for the risers, what combinations of step height and tread width satisfy the rule of thumb if you can only use integer lengths? (There are 8 possible combinations.)

5. Which of the combinations from exercise 4 has the steepest slope? Which has the flattest slope?

Suppose the stairs are built with a step height of 5 inches and a tread width of 14 inches, and the thickness of the step is 1 inch. How many steps would it take to reach a second story that is ten feet above the first story?

Each step is 1 inch thick, adding that to the 5-inch step height produces a rise of 6 inches per step. Since 10 feet is equal to 120 inches, it will take 20 steps $(120 \div 6 = 20)$ to reach the second story.

Each step has a tread width of 14 inches, so 20 steps would take up a horizontal distance of $20 \cdot 14$ or 280 inches. That is a horizontal distance of 23 feet and 4 inches.

---

6. For the flattest slope from exercise 5, how many steps would it take to reach a second story ten feet up? What is the horizontal distance covered by the staircase?
(Assume your individual steps are each one inch thick, and all joints fit perfectly.)

7. For the steepest slope from exercise 5, how many steps would it take to reach a second story ten feet up? What is the horizontal distance covered by the staircase?
(Assume your individual steps are each one inch thick, and all joints fit perfectly.)

Car Depreciation

According to the car buying website Carfax.com, a new car tends to lose 20% in value after the first 12 months, and then 10% of its current value annually for the next two years, then 12% the next year.

Suppose you bought a new car for $40,000. Find the value of the car after the given number of years.

1. One year

2. Two years

3. Three years

4. Four years

5. Plot the ordered pairs for the initial value of the car as well as the value after the first 4 years, where x represents the number of years and y represents the value of the car.

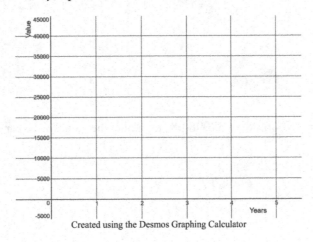

Created using the Desmos Graphing Calculator

6. Compute the slopes between each successive pair of points.

a. Slope for years 0 and 1	b. Slope for years 1 and 2	c. Slope for years 2 and 3	d. Slope for years 3 and 4

7. What is the cause of the different slopes in exercise 5?

The line that best fits the data, called a regression line, has the equation $y = -4046.08x + 37998.08$.

We can also create a quadratic model for the data using quadratic regression. The equation for the quadratic model is $y = 721.37x^2 - 6931.57x + 39440.82$. Here are the graphs of each model.

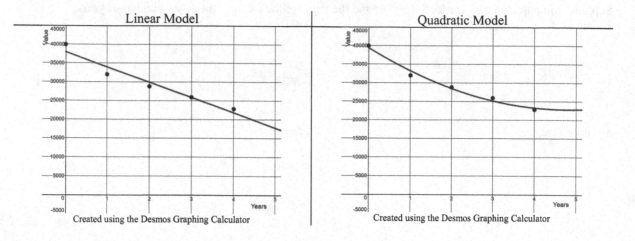

Created using the Desmos Graphing Calculator

8. Which model fits the data points better? Explain your answer.

9. Use the linear model, $y = -4046.08x + 37998.08$, to predict the value of the car when it is six years old.

10. Use the quadratic model, $y = 721.37x^2 - 6931.57x + 39440.82$, to predict the value of the car when it is six years old.

Loans

The monthly payment for an installment loan, like a car loan, can be computed using the formula

$$\text{Payment} = P \cdot \frac{r(1+r)^n}{(1+r)^n - 1}$$

where P is the principal, r is the interest rate per compounding period, and n is the number of compounding periods.

Suppose you wanted to take out a $50,000 car loan. If the annual interest rate is 4.9% and the length of the loan is 4 years, find the monthly payment.

The principal, P, is $50,000. The annual interest rate is 4.9%, so the rate per month is $\frac{0.049}{12}$. Interest is

compounded 12 times per year, so there are $4 \cdot 12$ or 48 compounding periods.

$$\text{Payment} = 50,000 \cdot \frac{\dfrac{0.049}{12}\left(1+\dfrac{0.049}{12}\right)^{48}}{\left(1+\dfrac{0.049}{12}\right)^{48} - 1} = \$1149.20$$

The total amount of the payments is $48 \cdot \$1149.20 = \$55,161.60$.
The total amount of interest paid is $\$55,161.60 - \$50,000 = \$5,161.60$.

1. Compute the monthly payment for a $24,000 car loan under the given conditions.
a. 1.9% annual interest, 5-year loan   b. 2.9% annual interest, 3-year loan

c. Which loan has less total interest?

2.

a. Compute the monthly mortgage payment for a $300,000 30-year mortgage, with a 4.0% annual interest rate (compounded monthly).

b. How much interest is paid over the life of the loan?

Interest on the average daily credit card balance P is compounded daily. The annual interest rate, r, is divided by 365 when computing compound interest. In a month with t days, the balance can be found using the formula

$$A = P \cdot \left(1 + \frac{r}{365}\right)^t$$

If a credit card has an average daily balance of $1000 in a month of 30 days, and the annual interest rate is 10%, find the new balance after 30 days.

$$A = 1000 \cdot \left(1 + \frac{0.10}{365}\right)^{30} = \$1008.25$$

3. A credit card company charges an annual interest rate of 15.99% on unpaid balances, compounded daily. If your average daily balance last month was $700, but you only make the minimum payment of $25, how much will your balance be in 31 days assuming you make no further charges?

4. A credit card company charges an annual interest rate of 18% on unpaid balances, compounded daily. If your average daily balance last month was $5000, but you only make the minimum payment of $100, how much will your balance be in 30 days assuming you make no further charges?

5. To fund a startup business at the start of April, Aurora borrowed $3500 on her credit card. The annual interest rate is 12%. She plans to pay back $500 per month until the balance is paid off.

So, in April, the average daily balance was $3500 and there are 30 days in that month. The balance after 30 days is

$$A = 3500 \cdot \left(1 + \frac{0.12}{365}\right)^{30} = \$3534.69$$

After making a $500 payment, the average daily balance for May will be $3034.69.

This first month's transactions can be recorded in a table.

MONTH	BEGINNING BALANCE	BALANCE AFTER INTEREST	PAYMENT	ENDING BALANCE
APRIL (30 DAYS)	$3500.00	$3534.69	$500.00	$3034.69

Fill in the table until the balance is paid back.

MONTH	BEGINNING BALANCE	BALANCE AFTER INTEREST	PAYMENT	ENDING BALANCE
APRIL (30 DAYS)	$3500.00	$3534.69	$500.00	$3034.69
MAY (31 DAYS)	$3034.69	$3065.77	$500.00	$2565.77
JUNE (30 DAYS)	$2565.77	$2591.20	$500.00	$2091.20
JULY (31 DAYS)	$2091.20	$2112.62	$500.00	$1612.62
AUGUST (31 DAYS)	$1612.62	$1629.14	$500.00	$1129.14
SEPTEMBER (30 DAYS)	$1129.14	$1140.33	$500.00	$640.33
OCTOBER (31 DAYS)	$640.33	$646.89	$500.00	$146.89
NOVEMBER (30 DAYS)	$146.89	$148.35	$148.35	$0.00

6.

a. What is the total amount that Aurora paid back for the loan?

$3648.35

b. What was the total amount of interest she paid?

$148.35

c. What percent of the $3500 loan is the total interest? Round to the nearest tenth of a percent.

4.2%

Marginal Tax Rates

The US has seven federal income tax brackets in 2019, with rates of 10%, 12%, 22%, 24%, 32%, 35%, and 37%. These are called marginal tax rates, because the rate only applies to the amount of your taxable income in that bracket. (Taxable income is what remains after all credits and deductions are subtracted from your income.) The federal tax brackets for single filers are listed below.

Income level	Tax rate
$0 - $9525	10%
$9526 - $38,700	12%
$38,701 - $82,500	22%
$82,501 – $157,500	24%
$157,501 - $200,000	32%
$200,001 - $500,000	35%
$500,001 and above	37%

Suppose your taxable income was $9000, which is in the first tax bracket. Since 10% of $9000 is $900, you would owe $900 in taxes.

1. For income levels of $9525 or below, what would the linear function be for your total tax? Use x for income level and y for taxes.

If your income is in the second bracket, say $15,000, then your total tax is computed in two stages.
- First: Compute 10% of the first $9525
- Second: Compute 12% of the amount over $9525.

$$0.10 \cdot 9525 + 0.12(15,000 - 9525)$$
$$= 0.10 \cdot 9525 + 0.12 \cdot 5475$$
$$= 952.50 + 657$$
$$= 1609.50$$

The tax for a taxable income of $15,000 is $1609.50.

The tax for a taxable income in the second bracket can be expressed as the linear function
$f(x) = 0.10(9525) + 0.12(x - 9525)$. This form of this function can be simplified:
$$f(x) = 0.10(9525) + 0.12(x - 9525)$$
$$f(x) = 952.50 + 0.12x - 1143$$
$$f(x) = 0.12x - 190.50$$

We now have two parts of the tax function: one part for taxable incomes up to $9525 and a second part for taxable incomes between $9526 and $38,700. This type of function is called a piecewise function because it has different definitions on different intervals. The piecewise function for the first two brackets can be expressed as follows:

$$f(x) = \begin{cases} 0.10x & 0 \le x \le 9525 \\ 0.12x - 190.50 & 9526 \le x \le 38,700 \end{cases}$$

If a taxable income is $9525 or less, use $0.10x$ to compute the tax owed.
If a taxable income is anywhere from $9526 to $$38,700, use $0.12x - 190.50$ to compute the tax owed.

2. Use the piecewise function $f(x) = \begin{cases} 0.10x & 0 \le x \le 9525 \\ 0.12x - 190.50 & 9526 \le x \le 38,700 \end{cases}$ to compute the tax owed on a taxable income of $38,700.

3. If a person has a taxable income of $38,700, what percent of their income is the tax they owe? Round to the nearest tenth of a percent.

4. Complete the piecewise function for the first three brackets.

$$f(x) = \begin{cases} 0.10x & 0 \le x \le 9525 \\ 0.12x - 190.50 & 9526 \le x \le 38,700 \\ & 38,701 \le x \le 82,500 \end{cases}$$

5. Use the piecewise function to compute the tax owed on a taxable income of $82,500.

6. If a person has a taxable income of $82,500, what percent of their income is the tax they owe? Round to the nearest tenth of a percent.

7. Complete the piecewise function for the last four brackets.

$$f(x) = \begin{cases} 0.10x & 0 \leq x \leq 9525 \\ 0.12x - 190.50 & 9526 \leq x \leq 38,700 \\ 0.22x - 4060.50 & 38,701 \leq x \leq 82,500 \\ & 82,501 \leq x \leq 157,500 \\ & 157,501 \leq x \leq 200,000 \\ & 200,001 \leq x \leq 500,000 \\ & 500,001 \leq x \end{cases}$$

8. Use the piecewise function to compute the tax owed on the given taxable income, and determine what the effective tax rate (100% times tax divided by taxable income) is for that amount. (Round the effective tax rate to the nearest tenth of a percent.

a. $157,500

b. $200,000

c. $500,000

d. $1,000,000

Applying Statistical Summaries: Commuting to Campus

The thirty-two students in a statistics class took a survey to gather data about their commute to campus.

The circle graph below summarizes their primary modes of transportation.

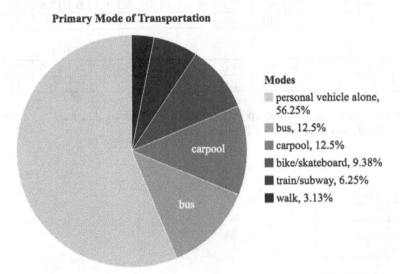

Primary Mode of Transportation

Modes
- personal vehicle alone, 56.25%
- bus, 12.5%
- carpool, 12.5%
- bike/skateboard, 9.38%
- train/subway, 6.25%
- walk, 3.13%

1. Create a bar graph for this data, with the vertical axis representing the number, not percent, of students in each category. Order the bars from tallest to shortest, thus creating a bar graph called a Pareto Chart.

2. To explore a relationship between distance and time, students consulted a phone app that calculated the driving distance, in miles, from their home to campus and also predicted the time, in minutes, it would take to drive that distance, given the current traffic conditions. The data for each student is listed column-wise in the tables below.

mi	2.0	9.5	12.9	5.6	18.2	6.2	17.1	5.4	14.8	6.3	22.0	17.8	9.1	21.3	7.0	24.8
min	3	13	16	10	22	12	26	8	20	9	31	21	11	32	9	31

mi	8.9	10.3	3.2	31.2	5.3	8.0	13.0	2.8	34.0	16.0	7.4	42.9	7.2	8.4	19.1	12.6
min	13	16	5	41	8	10	20	3	51	23	11	64	11	10	29	20

a. Construct a frequency distribution and histogram for the driving distances. Use 0 miles as the first lower class limit, and class width of 5 miles.

Distance in miles	Frequency
0 − 4.9	
5 − 9.9	
10 − 14.9	
15 − 19.9	
20 − 24.9	
25 − 29.9	
30 − 34.9	
35 − 39.9	
40 − 44.9	

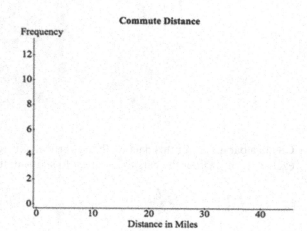

b. Given the shape of the distribution, predict whether the mean driving distance is greater than, less than, or approximately equal to the median driving distance. Calculate the mean and median to confirm your prediction.

c. Construct a relative frequency distribution for the driving time, in minutes.
Use 0 minutes as the first lower class limit and a class width of 5 minutes.
Insert a column for class midpoints and frequency.
(The class midpoint is equal to the average of that class's lower limit and the next class's lower limit.)
Round relative frequencies to three decimal places.

Time in minutes	Class Midpoint (min)	Frequency	Relative Frequency
0 − 4.9			
5 − 9.9			
10 − 14.9			
15 − 19.9			
20 − 24.9			
25 − 29.9			
30 − 34.9			
35 − 39.9			
40 − 44.9			
45 − 49.9			
50 − 54.9			
55 − 59.9			
60 − 64.9			

d. Theoretically, what it the sum of the relative frequencies in a relative frequency distribution? _____
What is the sum of the relative frequencies in the distribution above? _____. Why is it not equal to the theoretical sum?

e. Approximate the mean driving time by calculating the weighted mean of the class midpoints, using the frequencies as weights.

3. Do you expect there to be a correlation between commute distance and travel time? If so, how would this visible in a scatterplot? Explain.

4.   Complete the scatterplot for the paired data. Title the plot and the axes.

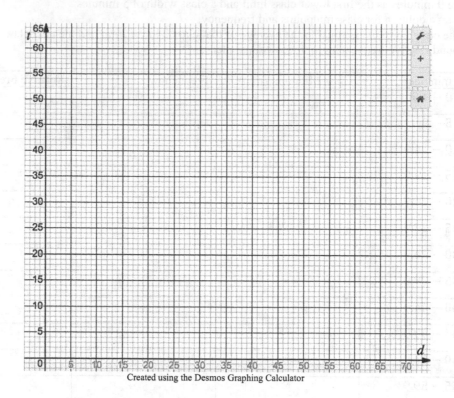

Created using the Desmos Graphing Calculator

a.   Why do the points not form a straight line?

b.   Draw an approximation to the line of best given by graphing the equation $t = \frac{7}{5}d$. Does the line appear to fit the data well enough to justify using the equation to make predictions?

c.   Use the equation $t = \frac{7}{5}d$ to predict the commute time for someone who lives 35 miles from campus.

d.   Use the equation $t = \frac{7}{5}d$ to predict the distance a student's home is from campus if his commute time is 14 minutes.

Applying Probabilities: Riddleys, Disbelieve Him or Not

A joker named Riddleys makes some outlandish statements about things that have happened to him by random chance. Some of his statements, though unlikely, might have actually happened. On the other hand, you decide to draw the line and flat-out disbelieve him if the probability of the event occurring by random chance is less than 0.05.

Calculate the probability of each event occurring by random chance. Round all probabilities to three decimal places. Based on the probability, state whether you disbelieve him or not. (To be clear, choosing not to disbelieve him does not mean you're saying you believe him. There's some middle ground for "maybe, maybe not.")

1. "I once flipped a fair coin and it landed heads-up twenty times in a row."

   a. Use the Fundamental Counting Rule to determine an expression for the number of different sequences of twenty heads and/or tails. (Hint: How many equally likely outcomes are there for the first flip? For the second? The third? And so on.)

   b. How many of the results consist of twenty heads?

   c. Compute the probability that the event described happened by random chance.

   d. Riddleys: <u>Disbelieve him</u> or <u>not</u>? (Circle one.)

2. "I hit "Shuffle" on my music app to randomize my "Chillin' Out" playlist, and it played all 12 of my songs in alphabetical order."

   a. How many arrangements are there of 12 distinct items?

   b. How many of those arrangements are alphabetical?

   c. Compute the probability that the event described happened by random chance.

   d. Riddleys: <u>Disbelieve him</u> or <u>not</u>? (Circle one.)

3. "I correctly guessed the month of the year in which my soccer coach was born."

   a. Compute the probability that the event described happened by random chance.

   b. Riddleys: <u>Disbelieve him</u> or <u>not</u>? (Circle one.)

4. "I have five nieces whose names are Alice, Beatrix, Cassie, Darla, Ellie. My two nephews' names are Simon and Garrett. I randomly selected one of my nieces, and I randomly selected one of my nephews, both to join me at a game at the stadium. I chose the youngest niece and the elder nephew."

   a. Draw a tree diagram to illustrate all of the different possible niece-nephew combinations.

   b. Compute the probability that the event described happened by random chance.

   c. Riddleys: <u>Disbelieve him</u> or <u>not</u>? (Circle one.)

5. "I once randomly selected four out of my seven nieces and nephews to ride in my car with my wife and me to the family reunion. It turns out I picked the four oldest of them."

   a. How many different sets of four of his nieces and nephews are possible?

   b. How many of those sets consist of the four oldest?

   c. Compute the probability that the event described happened by random chance.

   d. Riddleys: <u>Disbelieve him</u> or <u>not</u>? (Circle one.)

6. "My favorite fast-food restaurant sponsored a scratch-off bingo game. The odds in favor of winning a free order of fries was 3:47. Guess what? I won a free order of fries!"

   a. Compute the probability that the event described happened by random chance.

   b. Riddleys: <u>Disbelieve him</u> or <u>not</u>? (Circle one.)

7. "I didn't study for my statistics quiz, so I randomly guessed on three multiple-choice questions with five choices each and two true-false questions. I got 100%."

   a. Compute the probability that the event described happened by random chance.

   b. Riddleys: <u>Disbelieve him</u> or <u>not</u>? (Circle one.)

Venn Diagrams and Euler Diagrams

1. Match each of the following statements to the correct Euler diagram:
   a.   All mathematics professors are cool people.   _____
   b.   All cool people are mathematics professors   _____
   c.   Some cool people are mathematics professors   _____
   d.   Some cool people are not mathematics professors   _____

1

Cool people

Mathematics professors

2

mathematics professors

cool people

3

cool people

●

mathematics professors

4

mathematics professors

●

cool people

2. Determine if the argument is valid. (Use an Euler diagram to help make your decision.)
a. All vegetarians like oatmeal. Suzi likes oatmeal. Therefore, Suzi is a vegetarian.

b. All humans have a heart. Yolotli is a human. Therefore, Yolotli has a heart.

c. Some college classes are math classes. Jorge is taking a college class. Therefore, Jorge is taking a math class.

d. Some birds can fly. A penguin is a bird. Therefore, a penguin can fly.

3. If there are 100 total people, match each of the following statements to the correct Venn diagram:
    a.   All mathematics professors are cool people.   _____
    b.   All cool people are mathematics professors   _____
    c.   Some mathematics professors are cool people   _____
    d.   No cool people are mathematics professors   _____

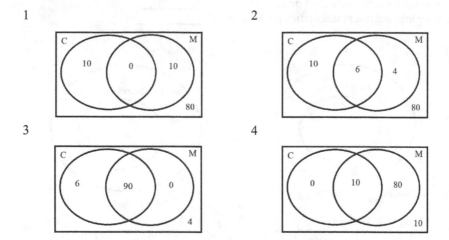

4. Match the statements with the correct Venn diagram(s): Let M=mathematics professors, C = cool people, and S = students. (So more than one diagram may match more than one statement)
    a.   All mathematics professors are cool people and all cool people are students.   _____
    b.   All mathematics professors are students and all students are cool people.   _____
    c.   All mathematics professors are cool people and some cool people are students.   _____
    d.   All mathematics professors are students and some students are cool people.   _____

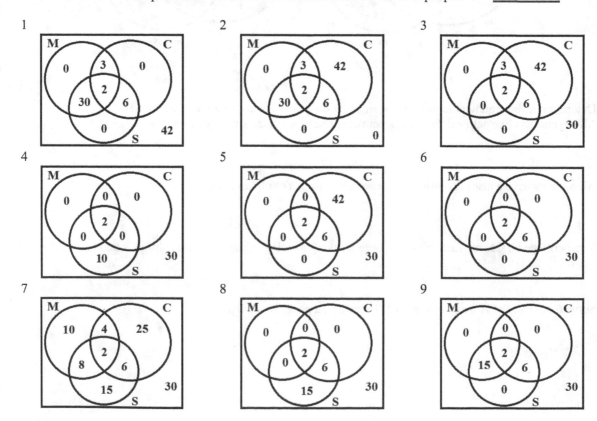